Standards for Technological Literacy:

Content for the Study of Technology

International Technology Education Association

and its

Technology for 

All Americans Project

This material is based upon work supported by the following:

National Science Foundation under Grant No. ESI-9626809 and the National Aeronautics and Space Administration under Grant No. NCC5-172. Any opinions, findings, and conclusions or recommendations expressed in this material are those of the author(s) and do not necessarily reflect the views of the National Science Foundation or the National Aeronautics and Space Administration.

ISBN: 1-887101-02-0

Copies of this document are being disseminated by the
International Technology Education Association
1914 Association Drive, Suite 201
Reston, Virginia 20191-1539
Phone: (703) 860-2100
Fax: (703) 860-0353
Email: itea@iris.org
URL: http://www.iteawww.org

Contents

Dedication

This publication is dedicated to two individuals who played a significant role in creating this project and then assisting with its early stages—Dr. Walter B. Waetjen (1920–1997) and Mr. Thomas A. Hughes, Jr. The project was launched because of their dedication and proceeded because of their enthusiasm to advance the teaching of technology. Their spirit and pursuit of excellence have served as guides in the creation of these standards for technology.

Foreword

We are a nation increasingly dependent on technology. Yet, in spite of this dependence, U.S. society is largely ignorant of the history and fundamental nature of the technology that sustains it. The result is a public that is disengaged from the decisions that are helping shape its technological future. In a country founded on democratic principles, this is a dangerous situation.

Thankfully, in *Standards for Technological Literacy: Content for the Study of Technology (Technology Content Standards)*, we have a tool to help us address the mismatch between dependency and understanding. Through an arduous four-year process, involving many levels of review and countless revisions, the International Technology Education Association has successfully distilled an essential core of technological knowledge and skills we might wish all K-12 students to acquire.

It is worth noting that the view of technological literacy spelled out in these standards includes reference to computers and the Internet, but it correctly does not focus unduly on these technologies, which comprise only a small part of our vast human-built world.

The standards and associated benchmarks in this document have been carefully written to ensure they are age appropriate. They are crafted to build increasingly sophisticated understanding and ability as students mature. In this way, *Technology Content Standards* provides an ambitious framework for guiding student learning.

The standards should not be viewed as static and immutable. Rather, *Technology Content Standards* — as is true for all good standards — will undergo periodic reassessment and reevaluation. It is very much a living document.

It is not enough that the standards are published. To have an impact, they must influence what happens in every K-12 classroom in America. This will not happen without the development of new curricula, textbooks, and student assessments, to name just a few of the more important factors. And, certainly, it cannot happen without the participation of teachers — all teachers, not just technology educators.

Indeed, the standards cannot succeed without the concerted effort of many stakeholder groups. In this regard, I urge all readers to review Chapter 8, Call to Action. As anyone involved in U.S. education knows, meaningful and lasting change occurs over many years, if not decades. While we need to be aware of this long timeline, we should not be discouraged by it. There is much to be done and much to be hopeful about. The ITEA standards provide a clear vision for the many individuals and organizations around the country committed to enhancing the technological literacy of the nation.

Wm. A. Wulf
President
National Academy of Engineering

Preface

With the growing importance of technology to our society, it is vital that students receive an education that emphasizes technological literacy. *Standards for Technological Literacy: Content for the Study of Technology* (referred to henceforth as *Technology Content Standards*) presents a vision of what students should know and be able to do in order to be technologically literate. These standards do not attempt to define a curriculum for the study of technology; that is something best left to states and provinces, school districts, and teachers. Instead, as the name implies, the standards describe what the content of technology education should be in grades K-12. By setting forth a consistent content for technology education in schools around the country, *Technology Content Standards* will help ensure that all students receive effective instruction about technology.

Technology Content Standards was created under the aegis of the International Technology Education Association and its Technology for All Americans Project (see Appendix A), and hundreds of educators and professionals have participated in its development and revision. We thank everyone who was involved in this important consensus-building process. As a result of their efforts, we believe that *Technology Content Standards* can be a catalyst for reform, bringing about significant change in the study of technology and resulting in the recognition of technology education as an essential core field of study in the schools.

Anyone interested in seeing that students receive a high-quality and relevant education, particularly those involved in decisions about what our schools teach, should find this document useful. The document's intended audience includes teachers, curriculum developers, school administrators, teacher educators, school board members, parents, engineers, business leaders, and others in the educational community, as well as the community as a whole.

Technology Content Standards does not represent an end, but a beginning. In other fields of study, developing standards has often proved to be the easiest step in a long, arduous process. Therefore, we can predict that getting these technology standards accepted and implemented in grades K-12 in every school will be far more difficult than developing them has been. Only through the combined efforts of educational decision-makers everywhere will we be able to ensure that all students develop higher levels of technological literacy.

This work has been made possible by the generous support of the National Science Foundation (NSF) and the National Aeronautics and Space Administration (NASA), and we would like to express our appreciation to both agencies. We are excited about the difference that *Technology Content Standards* can make, and we urge each of you to work collaboratively to use it as a basis for improving the study of technology.

William E. Dugger, Jr.
Director
Technology for All Americans Project
International Technology Education Association

Anthony F. Gilberti
President
International Technology Education Association

1 Preparing Students for a Technological World

Humans have been called the animals which make things, and at no time in history has that been so apparent as the present. Today, every human activity is dependent upon various tools, machines, and systems, from growing food and providing shelter to communication, healthcare, and entertainment. Some machines, like the tractor, speed up and make more efficient activities that humans have done for hundreds or thousands of years. Others, such as the airplane or the Internet, make possible things that humans have never been able to do before. This collection of devices, capabilities, and the knowledge that accompanies them is called technology.

1 Preparing Students for a Technological World

Broadly speaking, technology is how people modify the natural world to suit their own purposes. From the Greek word *techne*, meaning art or artifice or craft, technology literally means the act of making or crafting, but more generally it refers to the diverse collection of processes and knowledge that people use to extend human abilities and to satisfy human needs and wants.

The Need for Technological Literacy

Technology has been going on since humans first formed a blade from a piece of flint, harnessed fire, or dragged a sharp stick across the ground to create a furrow for planting seeds, but today it exists to a degree unprecedented in history. Planes, trains, and automobiles carry people and cargo from place to place at high speeds. Telephones, television, and computer networks help people communicate with others across the street or around the world. Medical technologies, from vaccines to magnetic resonance imaging, allow people to live longer, healthier lives. Furthermore, technology is evolving at an extraordinary rate, with new technologies being created and existing technologies being improved and extended.

All this makes it particularly important that people understand and are comfortable with the concepts and workings of modern technology. From a personal standpoint, people benefit both at work and at home by being able to choose the best products for their purposes, to operate the products properly, and to troubleshoot them when something goes wrong. And from a societal standpoint, an informed citizenry improves the chances that decisions about the use of technology will be made rationally and responsibly.

For these reasons and others, in the past several years, a growing number of voices have called for the study of technology to be included as a core field of study in elementary, middle, and secondary schools. Among the experts who have addressed the issue, the value and importance of teaching about technology is widely accepted.

Despite this consensus, however, technology laboratory-classrooms, the formal environment in the school where the study of technology takes place, are available in only a small number of elementary, middle, and secondary schools around the country. A few school districts have put comprehensive technology programs in place, and a handful of states and provinces have set forth technology standards, but nationwide most students receive little or no formal exposure to the study of technology. They are graduating with only a minimal understanding of one of the most powerful forces shaping society today.

The reasons for this situation are not hard to find. One is simple inertia. To keep doing what one has been doing is always easier than learning to do something new. A bigger reason, though, lies with the pressures on the educational system today. The back-to-basics push has emphasized competency in such traditional courses as English, mathematics, science, history and social studies, but technology has never been a basic part of education for most students. Furthermore, the growing emphasis on standardized competency tests has encouraged schools to teach to those tests, which generally contain few questions gauging technological literacy. So, squeezed for time and resources, relatively few local school districts and states or provinces have opted for what they see as the luxury of including the study of technology as part of the core curriculum.

Compounding these problems is the fact that the study of technology (technology education) is a mystery to many teachers and administrators. As a field of study that has evolved over the past fifteen to twenty years from industrial arts programs, technology education is just beginning to establish a new identity that people outside the field recognize and understand. There is still widespread confusion about the differences between technology education and educational technology, which uses technology as a tool to enhance the teaching and learning process.

The standards and enabling benchmarks in this document have been developed to help clear up this confusion and to build the case for technological literacy by setting forth precisely what the outcomes of the study of technology should be. Technology teachers, as well as science and mathematics teachers, and other educators and experts from around the country, collaborated to spell out what students in kindergarten through twelfth grade should be learning about

technology. These people, along with curriculum specialists and representatives from the National Research Council and the National Academy of Engineering, reviewed *Technology Content Standards* and suggested changes and additions. The result is a document that both defines the study of technology as a field of study and provides a road map for individual teachers, schools, school districts, and states or provinces that want to develop technological literacy in all students.

The standards presented here do more than provide a checklist for the technological facts, concepts, and capabilities that students should master at each level. Along the way, they explain how and why technological literacy fits with the broad mission of schools, and they describe the benefits of the study of technology for students. In short, they make the case for why — despite inertia, despite the back-to-basics movement, despite the growing emphasis on standardized competency exams, and despite the various other pressures on educators — the study of technology should be an integral part of the curriculum of our schools.

Learning About Technology

Students who study technology learn about the technological world that inventors, engineers, and other innovators have created. They study how energy is generated from coal, natural gas, nuclear power, solar power, and wind, and how it is transmitted and distributed. They examine communications systems: telephone, radio and television, satellite communications, fiber optics, the Internet. They delve into the various manufacturing and materials-processing industries, from steel and petrochemicals to computer chips and household appliances. They investigate transportation, information processing, and medical technology. They even look into new technologies, such as genetic engineering or emerging technologies, such as fusion power that is still years or decades away.

Because technology is so fluid, teachers of technology tend to spend less time on specific details and more on concepts and principles. The goal is to produce students with a more conceptual understanding of technology and its place in society, who can thus grasp and evaluate new bits of technology that they might never have seen before.

To this end, *Technology Content Standards* emphasizes comprehension of the basic elements that go into any technology. One of these elements, for example, is the design process, the main approach that engineers, designers, and others in technology use to create solutions to problems. Another is development and production, whereby the design is transformed into a finished product, and a system is created to produce it. A third element is the use and maintenance of the product, which can determine the product's success or failure. Each of these steps in the technological process demands its own set of skills and mental tools.

Besides understanding how particular technologies are developed and used, students should be able to evaluate their effects on other technologies, on the environment, and on society itself. The benefits of a technology are usually obvious — if they were not, it would probably never be developed — but the disadvantages and dangers are often hidden. When chlorofluorocarbons (CFCs) were invented, for example, no one realized that these chemicals used as refrigerants and blowing agents for foam would eventually damage the ozone layer. Today, the Internet is having profound effects on society — how people interact and communicate with one another, how they do business, and how they get their entertainment and recreation — but no one knows exactly what to expect from it in the long run.

One of the basic lessons in studying technology is that not only can technology be used to solve problems, but it may also create new ones. Many of these new problems can be solved or ameliorated by yet more technology, but this may in turn beget other problems, and so on. Technologies inevitably involve trade-offs between benefits and costs. Intelligent decisions made about a technology need to take both into account. Students should come to see each technology as neither good nor bad in itself, but one whose costs and benefits should be weighed to decide if it is worth developing.

Learning to Do Technology

One of the great benefits of learning about technology is also learning to do technology, that is, to carry out in the laboratory-classroom many of the processes that underlie the development of technology in the real world. Recent research on learning finds that many students learn best in experiential ways — by doing, rather than only by seeing or hearing — and the study of technology emphasizes and capitalizes on such active learning.

For instance, students in technology laboratory-classrooms are taught practical problem-solving skills and are asked to put them to work on different types of real-world problems. Engineers, architects, computer scientists, technicians, and others involved in technology use a variety of approaches to problem solving, including troubleshooting, research and development, invention, innovation, and experimentation. Students will become familiar with these approaches and learn about the appropriate situations in which to use them. They will also learn that design (sometimes called "technological design") is the primary problem-solving approach in technology. In learning to design, students will master a set of abilities that will serve them well throughout their lives.

The design process generally begins with identifying and defining a problem — there

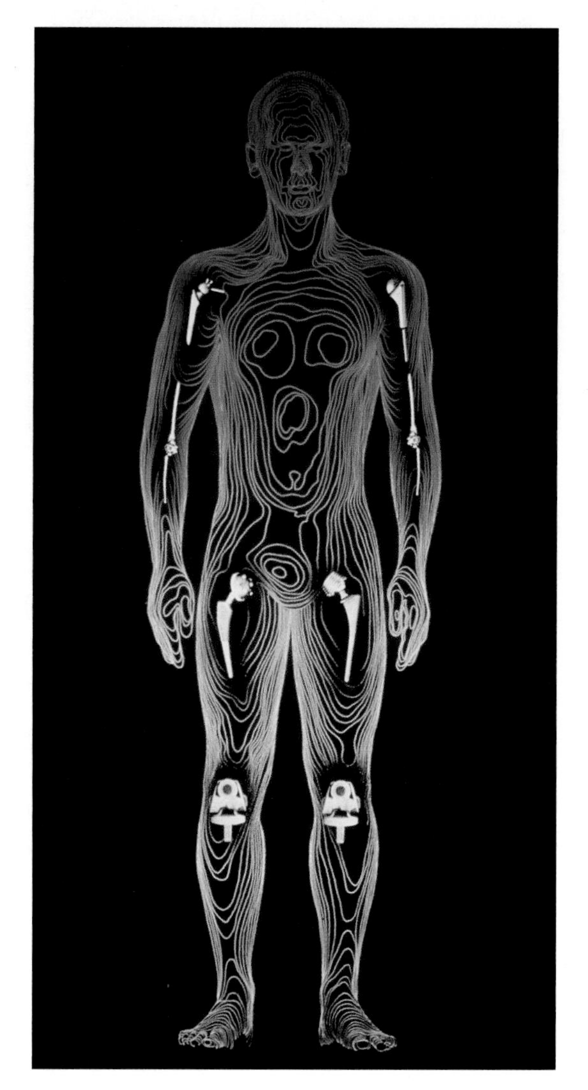

is some need to be met or some want to be fulfilled, and the designer must understand exactly what it is. After investigating and researching the problem, the designer generates a number of ideas for a solution. Because it is particularly helpful for several people to brainstorm ideas, students will generally work in groups at this stage. Then, considering the original criteria, along with various constraints, one design — or, in some cases, more than one — is chosen as the most promising. The selected design is modeled and tested, and then reevaluated. If necessary, the original design is dropped and another is tried. Eventually, through a series of iterations, repeating the various steps of the process as necessary, a final design is chosen.

This design process can be applied to almost any sort of design. In one elementary school classroom, for instance, the students were asked to design and build a "pop rocket" to demonstrate Newton's Third Law. In high school technology laboratory-classrooms, one assignment might be to design a water-purification system for a catfish farm. One of the first lessons that students learn from exercises like these is that there are many possible solutions to a technological problem, and that while some answers are clearly wrong — they don't work, or they work poorly — there is no such thing as "the" correct answer.

Such design projects are inevitably more than just mental exercises. Students generally work in teams when building models of their design proposals, and, depending on the device, they may build working prototypes as well. Such hands-on learning engages the students in a way that lectures, problem solving on paper, or lab exercises that follow a preset series of steps cannot. In other words, design exercises encourage active learning rather than passive learning.

In addition to problem-solving skills, students are given opportunities to use and maintain technological products correctly, again with an emphasis on learning how to learn. Because it would be impossible to instruct students on every product they might encounter, they are given experience with some common tools and systems to gain familiarity with the basic principles of using and maintaining technological products. They are also taught how to learn about products on their own — by reading instructions, or searching for information on the Internet, for instance. The confidence and familiarity with technology that they acquire will prepare students to deal intelligently with current and future technological products.

Technological Studies as an Integrator

Perhaps the most surprising message to emerge from *Technology Content Standards* — surprising, at least, to those who have not themselves taught technology classes — is the role technological studies can play in students' learning of other subjects. When taught effectively, technology is not simply one more field of study seeking admission to an already crowded curriculum, pushing others out of the way. Instead, it reinforces and complements the material that students learn in other classes.

As envisioned by the standards in the following chapters, the study of technology is a way to apply and integrate knowledge from many other subject areas — not just

Technology is the modification of the natural environment in order to satisfy perceived human needs and wants.

mathematics, science, and computer classes, but also the liberal and fine arts. Consider, for instance, a field trip taken by a class of fourth-graders in Michigan to Greenfield Village, a historical museum with restored houses and shops. The class had just finished a history unit on America at the turn of the twentieth century, which prepared them for what they would find. While there, each class member chose an artifact of a particular technology used at the time — a hay thresher, for instance, or a light bulb, or a car, or a clothes washing machine — and acted as a reporter by quizzing the docents for details about that device. Later, each student sketched and determined the proper scale needed to make a model of the artifact he or she had chosen. The students made models and prepared reports on the devices, including such information as their purpose, how they were made, how they were used, their roles in the economic and social life of the village, and a description of how they worked. Afterward, the class worked together to develop and create a video that would describe the technology of Greenfield Village as a part of a communications technology unit for future fourth-grade classes. The assignment taught the students a good deal about the technology of the era, and it reinforced lessons from other parts of the curriculum. The assignment brought turn-of-the-century America to life in history class; it exercised composition skills from English class; and it allowed the students to apply what they had learned in a motion and forces unit in science class. As teachers all know, having students apply material in a way that captures their interest and imagination is the best way to make sure they retain it. And when students can bring together lessons from several classes or content areas, they truly make the material their own.

Technological literacy is the ability to use, manage, assess, and understand technology.

Such integration among subjects is easiest in elementary schools where the same teacher

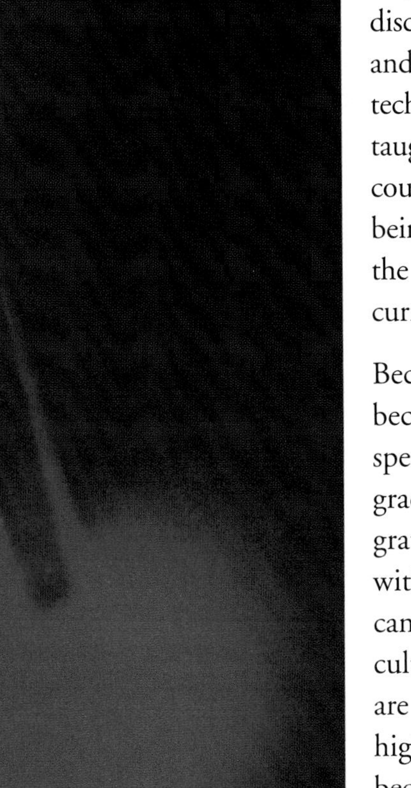

handles most or all of a student's classes during the school day and does not have to work with several other teachers to coordinate lesson plans. At the elementary level, the standards are designed to be implemented in the regular classroom by teachers with appropriate in-service training. In middle and high schools, by contrast, licensed technology teachers should work with others in the delivery of technological studies; in the middle grades, much of the teaching about technology can be done in units taught by interdisciplinary teams; and in high school technology should be taught in stand-alone courses as well as being integrated into the rest of the curriculum.

Because instruction becomes increasingly specialized at higher grade levels, integrating technology with other subjects can be more difficult, but the payoffs are proportionately higher. As subjects become more compartmentalized, students find it more difficult to see how they intersect with one another or to understand the place of each in the world outside. Technology laboratory-classrooms provide a neutral ground for different subjects to come together, often in the guise of devising a solution to some practical problem. A typical assignment might be to design a car with certain characteristics — being crash-worthy, energy efficient, or using alternative fuels. In developing their designs, students could operate various computer programs and perhaps retrieve information from the Internet, apply lessons from physics or chemistry classes, and use skills from their mathematics classes as well. In researching the background to their problem, they might delve into the history of the car and how it has shaped American society in the twentieth century. They might use statistics to analyze automobile fatality rates at different speeds and in cars of various sizes. They could study the chemistry or the health effects of the ozone smog afflicting cities, or they could analyze the economics of gasoline prices. In an attempt to understand the world's petroleum reserves, they might study geology and explore how petroleum is formed. When writing a report on their final product, they would need to do so in clear prose, probably with a bibliography. They might even translate it into a second language or convert it into 'HTML' format for access on the Internet.

Many teachers have reported that this sort of real-world problem solving helps students with their other courses by making the subject matter meaningful to them. The best way to learn something — to truly master and retain it, not just to learn it well enough to pass a test — is to apply it. This, of course, is the rationale for lab sessions in chemistry class, word problems in mathematics, and conversational periods in French. But technology classes take this logic one step further because students are expected to synthesize and apply information from other subjects as well as from within the study of technology. In this way, they learn to make connections among different fields of study and begin to understand how all knowledge is interconnected.

People who are unfamiliar with technology tend to think of it purely in terms of its artifacts: computers, cars, televisions, toasters, pesticides, flu shots, solar cells, genetically engineered tomatoes, and all the rest. But to its practitioners and to the people who study it, technology is more accurately thought of in terms of the knowledge and the processes that create these products, and these processes are intimately dependent upon many factors in the outside world.

Technology is the modification of the natural environment in order to satisfy perceived human needs and wants. To determine what those needs and wants are and to figure out how to satisfy them, one must consider a wide range of factors simultaneously. For this reason, although the study of technology may sometimes be a separate subject, it can never be an isolated subject, cut off from the rest of the curriculum.

Technological Literacy

Technology Content Standards is designed as a guide for educating students in developing technological literacy. Technological literacy is the ability to use, manage, assess, and understand technology. A technologically literate person understands, in increasingly sophisticated ways that evolve over time, what technology is, how it is created, and how it shapes society, and in turn is shaped by society. He or she will be able to hear a story about technology on television or read it in the newspaper and evaluate the information in the story intelligently, put that information in context, and form an opinion based on that information. A technologically literate person will be comfortable with and objective about technology, neither scared of it nor infatuated with it.

Such technological literacy benefits students in a number of ways. For the future engineers, the aspiring architects, the students who will have jobs in one area of technology or another, it means they will leave high school with a head start on their careers. They will already understand the basics of such things as the design process, and they will have a big picture of the field they are entering, allowing them to put the specialized knowledge they learn later into a broader context.

But technological literacy is important for all students, even those who will not go into technological careers. Because technology is such an important force in our economy, anyone can benefit by being familiar with it. Corporate executives and others in the business world, brokers and investment analysts, journalists, teachers, doctors, nurses, farmers and homemakers all will be able to perform their jobs better if they are technologically literate.

On the individual level, technological literacy helps consumers better assess products and make more intelligent buying decisions: What are the important factors in evaluating the latest computer or electronic device? Should I avoid genetically engineered food? Should I put my infant in cloth or disposable diapers? A few years from now do I buy a solar-powered car or one that runs on hydrogen? Among people who have no familiarity with or basis for evaluating technological products, some of these decisions will be based simply on guesswork, gut feelings, or emotional responses.

On the societal level, technological literacy should also help citizens make better decisions. As the 21st century dawns, new technologies will open up possibilities for humankind that have never existed before. This power will bring with it hard choices. Do we place limits on the flow of information? How much heed do we pay to the worries that genetic engineering could lead to the inadvertent creation of unwelcome new species? At the same time, older, established technologies will also demand that choices be made: Should we, for instance, cut back sharply on carbon dioxide emissions in an attempt to slow down global warming?

In the United States, such decisions will be greatly influenced by individual citizens. In some countries, average citizens have little input into technological decision making, which is left to a technological elite or the country's rulers. But the political structure of the United States is open, and regular citizens can — and generally do — shape technological issues through their legislators, through public hearings, and through court cases. Having a technologically literate citizenry may not guarantee that the best decisions are made on these knotty, contentious issues, but it certainly improves the odds.

Our world will be very different 10 or 20 years from now. We have no choice about that. We do, however, have a choice whether we march into that world with our eyes open, deciding for ourselves how we want it to be, or whether we let it push us along, ignorant and helpless to understand where we're going or why. Technological literacy will make a difference.

2 Overview of *Technology Content Standards*

Because of the power of today's technological processes, society and individuals need to decide what, how, and when to develop or use various technological systems. Since technological issues and problems have more than one available solution, decision-making should reflect the values of the people and help them reach their goals. Such decision-making depends upon all citizens, both individually and collectively, acquiring a basic level of technological literacy — the ability to use, manage, and understand technology.

Technology for All Americans:
A Rationale and Structure for the Study of Technology
1996

2 Overview of *Technology Content Standards*

Technology Content Standards* **presents, in a coherent manner, what students should know and be able to do in order to achieve a high level of technological literacy. In other words, the standards prescribe what the outcomes of the study of technology in grades K-12 should be, but they do not put forth a curriculum to achieve those outcomes.** ***Technology Content Standards*** **also is designed to act as a catalyst for educational reform.**

The foundation for *Technology Content Standards* was laid with the publication of *Technology for All Americans: A Rationale and Structure for the Study of Technology (Rationale and Structure).* The *Rationale and Structure* presents the structure and content for the study of technology, and *Technology Content Standards* is an extension and elaboration of that earlier work.

Technology Content Standards

Technology programs across the United States generally have varying structures and content. Thus, a student who takes a technology course in one area of the country may not receive the same core information or study the same basic concepts and principles as a student in another area, even when the course titles are the same or nearly so. *Technology Content Standards* offers a way to provide a consistent content for the study of technology that will enhance the learning of K-12 students no matter where they live and what their future goals may be.

In the following chapters, *Technology Content Standards* lays out what should be learned and accomplished by each student in the study of technology at four levels, beginning with grades K-2 and continuing through 3-5, 6-8, and 9-12. The standards and benchmarks are tailored to be age and gender appropriate and are planned so that the material at each level builds on, amplifies, and extends the standards and benchmarks of earlier grades. Furthermore, the standards and benchmarks have been designed to prescribe the content knowledge and abilities of what students should know and be able to do in order to be technologically literate.

Technology Content Standards ***lays out what should be learned and accomplished by each student in the study of technology at four levels.***

Features of *Technology Content Standards*

Technology Content Standards was created with the following basic features:

- It offers a common set of expectations for what students in technology laboratory-classrooms should learn.
- It is developmentally appropriate for students.
- It provides a basis for developing meaningful, relevant, and articulated curricula at the local, state, and provincial levels.
- It promotes content connections with other fields of study in grades K-12.

Technology Content Standards is *not* a curriculum. A curriculum provides the specific details on how the content is to be delivered, including organization, balance, and the various ways of presenting the content in the laboratory-classroom, while standards describe what the content should be. Curriculum developers, teachers, and others should use *Technology Content Standards* as a guide for developing appropriate curricula, but the standards do not specify what should go on in the laboratory-classroom.

In laying out the essentials for the study of technology, *Technology Content Standards* represents a recommendation from educators, engineers, scientists, mathematicians, and parents about what skills and knowledge are needed in order to become technologically literate. It is not, however, a federal policy or mandate.

Technology Content Standards does not prescribe an assessment process for determining how well students are meeting the standards, although it does provide criteria

for such an assessment. Assessment practices deal with how well students learn the content in *Technology Content Standards*. Closely tied with assessment is how well a teacher has directly taught and guided students in the learning process, as well as how much support the school and school district have provided in this effort. The ultimate goal in any educational assessment process is to be able to determine how well each student is attaining technological literacy in grades K-12. Assessment takes place in many forms, from daily records of students' work, interviews, quizzes, and tests, portfolios of longitudinal activities in the laboratory-classroom, to standardized tests administered by the school system or state. Plans for comprehensive assessment throughout the student's education must be designed, implemented, and continually monitored.

Benchmarks in **Technology Content Standards** *provide the fundamental content elements for the broadly stated standards.*

2

Format of *Technology Content Standards*

The individual standards presented in *Technology Content Standards* are organized into five major categories, each of which is addressed in a separate chapter:

1. The Nature of Technology (Chapter 3)
2. Technology and Society (Chapter 4)
3. Design (Chapter 5)
4. Abilities for a Technological World (Chapter 6)
5. The Designed World (Chapter 7)

Each chapter begins with a narrative that defines a category, explains the importance of each topic within a category, and gives a brief overview of the chapter. The standards in Chapter 3 ask that students understand what technology is, become familiar with its concepts, and recognize the relationships between technology and other fields of study. Chapter 4 examines the use of technology in a broader context by examining its effects on human society and the physical environment, by exploring how societal factors shape technology, and by tracing the history of technology. The standards in Chapter 5 focus on a cognitive understanding of a design process with an emphasis on the attributes of design, the engineering design process, and other problem-solving approaches. Chapter 6 deals with developing abilities in designing, making, developing, operating, maintaining, managing, and assessing technological products and systems. Chapter 7 covers selecting, using, and understanding major technologies that are common today. See Table 2.1 for a listing of the chapters and the standard topics.

Standards

Technology Content Standards specifies what every student should know and be able to do in order to be technologically literate, and it offers criteria to judge progress toward a vision of technological literacy for all students. There are a total of 20 standards in this document and the individual standards fall into two types: what students should know and understand about technology, and what they should be able to do. The first type, which could be termed "cognitive" standards, sets out basic knowledge about technology — how it works, and its place in the world — that students should have in order to be technologically literate. The second type, the "process" standards, describes the abilities that students should have. The two types of standards are complementary. For example, a student can be taught in a lecture about a design process, but the ability to actually use a design process and to apply it for finding a solution to a technological problem comes only with hands-on experiences. Likewise, it is difficult to perform a design process effectively without having some theoretical knowledge of how it is usually done. See Appendix B for a comprehensive listing of the standards.

Benchmarks

Benchmarks in *Technology Content Standards* provide the fundamental content elements for the broadly stated standards. Benchmarks, which are statements that provide the knowledge and abilities that enable students to meet a given standard, are provided for each of the 20 standards at the K-2, 3-5, 6-8, and 9-12 grade levels. The benchmarks are identified by an alphabetical listing (e.g., A, B, C) and are highlighted in bold type. They are followed by supporting sentences (not in bold) that provide further detail, clarity, and examples.

TABLE 2.1 Listing of Technology Content Standards

CHAPTERS	STANDARDS	
3 Students will develop an understanding of The Nature of Technology. This includes acquiring knowledge of:	1	The characteristics and scope of technology.
	2	The core concepts of technology.
	3	The relationships among technologies and the connections between technology and other fields.
4 Students will develop an understanding of Technology and Society. This includes learning about:	4	The cultural, social, economic, and political effects of technology.
	5	The effects of technology on the environment.
	6	The role of society in the development and use of technology.
	7	The influence of technology on history.
5 Students will develop an understanding of Design. This includes knowing about:	8	The attributes of design.
	9	Engineering design.
	10	The role of troubleshooting, research and development, invention and innovation, and experimentation in problem solving.
6 Students will develop Abilities for a Technological World. This includes becoming able to:	11	Apply the design process.
	12	Use and maintain technological products and systems.
	13	Assess the impact of products and systems.
7 Students will develop an understanding of The Designed World. This includes selecting and using:	14	Medical technologies.
	15	Agricultural and related biotechnologies.
	16	Energy and power technologies.
	17	Information and communication technologies.
	18	Transportation technologies.
	19	Manufacturing technologies.
	20	Construction technologies.

The benchmarks are required in order for students to meet the standards.

Research in education has shown that when previously learned knowledge is tapped and built on, students are more likely to acquire a more coherent and thorough understanding of these processes than if they are taught as isolated abstractions (National Research Council, 1999).

With this in mind, the benchmarks are articulated from grades K-2 through 9-12 to progress from very basic ideas at the early elementary school level to more complex and comprehensive ideas at the high school level. Certain content "concepts" are found in the benchmarks, which extend across various levels to ensure continual learning of an important topic related to a standard.

Vignettes

A selection of vignettes is included in this document to provide snapshots of laboratory-classroom experiences. They offer detailed examples of how the standards can be implemented by a teacher. A large majority of the vignettes are authentic in that they have been successfully used before in an actual laboratory-classroom with students. A few of the vignettes were generated especially for these standards and are fictional — they have not been tried and tested. Readers should be cautioned that any vignette should not be read too literally or narrowly, nor should they be interpreted as a curriculum.

Format

The format of each standard follows this structure: (See Figure 2.1 for a sample layout)

1. The standard is expressed in sentence form.
2. A narrative follows that explains the intent of the standard.
3. Grade-level material is presented next for grades K-2, 3-5, 6-8, and 9-12.
4. Under each grade-level, a narrative follows that explains the standard at the grade-level under discussion and provides suggestions on how the standard can be implemented in the laboratory-classroom.
5. Each grade-level essay is followed by a series of benchmarks in bold type that detail the particular knowledge and abilities that students must attain in order to meet the standard. Each benchmark is further explained by supporting sentences (not in bold type) that offer examples and additional details.
6. Vignettes, which are scattered throughout each of the chapters, provide examples of laboratory-classroom experiences and offer illustrations of how the standards can be put into practice.

The standards and benchmarks were established for guiding a student's progress toward technological literacy. References that were used in the development of *Technology Content Standards* include the following standards in other subject areas: *National Science Education Standards* (National Research Council, 1996); *Benchmarks for Science Literacy* (American Association for the Advancement of Science, 1993); *Curriculum and Evaluation Standards for School Mathematics* (National Council of Teachers of Mathematics, 1989); and *National Council of Teachers of Mathematics Standards 2000 Draft* (NCTM, 1998) and others.

FIGURE 2.1 Organizational Format of a Sample Standard and Benchmark

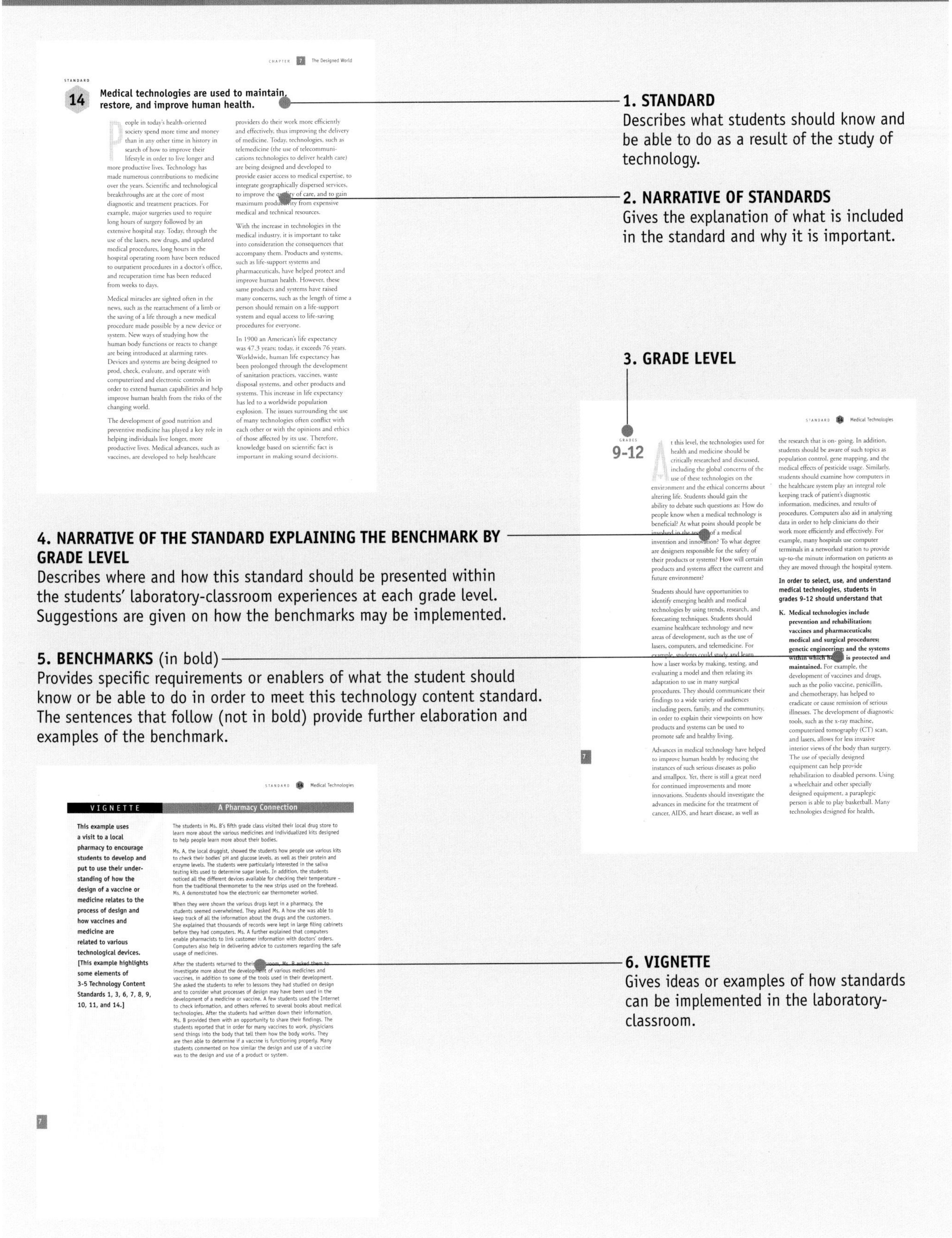

CHAPTER 7 The Designed World

STANDARD 14

Medical technologies are used to maintain, restore, and improve human health.

People in today's health-oriented society spend more time and money than in any other time in history in search of how to improve their lifestyle in order to live longer and more productive lives. Technology has made numerous contributions to medicine over the years. Scientific and technological breakthroughs are at the core of most diagnostic and treatment practices. For example, major surgeries used to require long hours of surgery followed by an extensive hospital stay. Today, through the use of the lasers, new drugs, and updated medical procedures, long hours in the hospital operating room have been reduced to outpatient procedures in a doctor's office, and recuperation time has been reduced from weeks to days.

Medical miracles are sighted often in the news, such as the reattachment of a limb or the saving of a life through a new medical procedure made possible by a new device or system. New ways of studying how the human body functions or reacts to change are being introduced at alarming rates. Devices and systems are being designed to prod, check, evaluate, and operate with computerized and electronic controls in order to extend human capabilities and help improve human health from the risks of the changing world.

The development of good nutrition and preventive medicine has played a key role in helping individuals live longer, more productive lives. Medical advances, such as vaccines, are developed to help healthcare providers do their work more efficiently and effectively, thus improving the delivery of medicine. Today, technologies, such as telemedicine (the use of telecommunications technologies to deliver health care) are being designed and developed to provide easier access to medical expertise, to integrate geographically dispersed services, to improve the quality of care, and to gain maximum productivity from expensive medical and technical resources.

With the increase in technologies in the medical industry, it is important to take into consideration the consequences that accompany them. Products and systems, such as life-support systems and pharmaceuticals, have helped protect and improve human health. However, these same products and systems have raised many concerns, such as the length of time a person should remain on a life-support system and equal access to life-saving procedures for everyone.

In 1900 an American's life expectancy was 47.3 years; today, it exceeds 76 years. Worldwide, human life expectancy has been prolonged through the development of sanitation practices, vaccines, waste disposal systems, and other products and systems. This increase in life expectancy has led to a worldwide population explosion. The issues surrounding the use of many technologies often conflict with each other or with the opinions and ethics of those affected by its use. Therefore, knowledge based on scientific fact is important in making sound decisions.

1. STANDARD
Describes what students should know and be able to do as a result of the study of technology.

2. NARRATIVE OF STANDARDS
Gives the explanation of what is included in the standard and why it is important.

3. GRADE LEVEL

STANDARD 14 Medical Technologies

GRADES 9-12

At this level, the technologies used for health and medicine should be critically researched and discussed, including the global concerns of the use of these technologies on the environment and the ethical concerns about altering life. Students should gain the ability to debate such questions as: How do people know when a medical technology is beneficial? At what point should people be involved in the testing of a medical invention and innovation? To what degree are designers responsible for the safety of their products or systems? How will certain products and systems affect the current and future environment?

Students should have opportunities to identify emerging health and medical technologies by using trends, research, and forecasting techniques. Students should examine healthcare technology and new areas of development, such as the use of lasers, computers, and telemedicine. For example, students could study and learn how a laser works by making, testing, and evaluating a model and then relating its adaptation to use in many surgical procedures. They should communicate their findings to a wide variety of audiences including peers, family, and the community, in order to explain their viewpoints on how products and systems can be used to promote safe and healthy living.

Advances in medical technology have helped to improve human health by reducing the instances of such serious diseases as polio and smallpox. Yet, there is still a great need for continued improvements and more innovations. Students should investigate the advances in medicine for the treatment of cancer, AIDS, and heart disease, as well as the research that is on- going. In addition, students should be aware of such topics as population control, gene mapping, and the medical effects of pesticide usage. Similarly, students should examine how computers in the healthcare system play an integral role keeping track of patient's diagnostic information, medicines, and results of procedures. Computers also aid in analyzing data in order to help clinicians do their work more efficiently and effectively. For example, many hospitals use computer terminals in a networked station to provide up-to-the minute information on patients as they are moved through the hospital system.

In order to select, use, and understand medical technologies, students in grades 9-12 should understand that

K. **Medical technologies include prevention and rehabilitation; vaccines and pharmaceuticals; medical and surgical procedures; genetic engineering; and the systems within which health is protected and maintained.** For example, the development of vaccines and drugs, such as the polio vaccine, penicillin, and chemotherapy, has helped to eradicate or cause remission of serious illnesses. The development of diagnostic tools, such as the x-ray machine, computerized tomography (CT) scan, and lasers, allows for less invasive interior views of the body than surgery. The use of specially designed equipment can help provide rehabilitation to disabled persons. Using a wheelchair and other specially designed equipment, a paraplegic person is able to play basketball. Many technologies designed for health,

7

4. NARRATIVE OF THE STANDARD EXPLAINING THE BENCHMARK BY GRADE LEVEL
Describes where and how this standard should be presented within the students' laboratory-classroom experiences at each grade level. Suggestions are given on how the benchmarks may be implemented.

5. BENCHMARKS (in bold)
Provides specific requirements or enablers of what the student should know or be able to do in order to meet this technology content standard. The sentences that follow (not in bold) provide further elaboration and examples of the benchmark.

STANDARD 14 Medical Technologies

VIGNETTE — A Pharmacy Connection

This example uses a visit to a local pharmacy to encourage students to develop and put to use their understanding of how the design of a vaccine or medicine relates to the process of design and how vaccines and medicine are related to various technological devices. [This example highlights some elements of 3-5 Technology Content Standards 1, 3, 6, 7, 8, 9, 10, 11, and 14.]

The students in Ms. B's fifth grade class visited their local drug store to learn more about the various medicines and individualized kits designed to help people learn more about their bodies.

Ms. A, the local druggist, showed the students how people use various kits to check their bodies' pH and glucose levels, as well as their protein and enzyme levels. The students were particularly interested in the saliva testing kits used to determine sugar levels. In addition, the students noticed all the different devices available for checking their temperature – from the traditional thermometer to the new strips used on the forehead. Ms. A demonstrated how the electronic ear thermometer worked.

When they were shown the various drugs kept in a pharmacy, the students seemed overwhelmed. They asked Ms. A how she was able to keep track of all the information about the drugs and the customers. She explained that thousands of records were kept in large filing cabinets before they had computers. Ms. A further explained that computers enable pharmacists to link customer information with doctors' orders. Computers also help in delivering advice to customers regarding the safe usage of medicines.

After the students returned to their classroom, Ms. B asked them to investigate more about the development of various medicines and vaccines, in addition to some of the tools used in their development. She asked the students to refer to lessons they had studied on design and to consider what processes of design may have been used in the development of a medicine or vaccine. A few students used the Internet to check information, and others referred to several books about medical technologies. After the students had written down their information, Ms. B provided them with an opportunity to share their findings. The students reported that in order for many vaccines to work, physicians send things into the body that tell them how the body works. They are then able to determine if a vaccine is functioning properly. Many students commented on how similar the design and use of a vaccine was to the design and use of a product or system.

7

6. VIGNETTE
Gives ideas or examples of how standards can be implemented in the laboratory-classroom.

Primary Users of *Technology Content Standards*

A variety of groups and individuals can be expected to use *Technology Content Standards*. Curriculum developers at the state, province, and local level, along with textbook publishers and developers of laboratory equipment, may be among the first users of the document. They will use it to fashion curriculum and resources for each grade-level.

Ultimately, of course, once the standards have been adopted at the local and state or provincial level, teachers will implement them. Some teachers may not wait for their states or provinces, or school districts to act. Either way, the ultimate success of *Technology Content Standards* rests with teachers.

Other users of *Technology Content Standards* will include superintendents, principals, and other administrators, curriculum coordinators, directors of instruction, and supervisors, all of whom will be a part of the planning, overseeing, and delivering of standards-based education. Teacher educators should use the document in designing pre-service programs for future technology teachers. Furthermore, parents should familiarize themselves with the document in order to become involved with their children's education and to reinforce the concepts and processes being taught. Finally, if students are home schooled, parents should incorporate *Technology Content Standards* into their instruction.

Recommendations for Using *Technology Content Standards*

Individuals involved in curriculum development, teaching, or assessment should consider the following recommendations:

- *Technology Content Standards* represents the careful thought of many people and is meant to be used in its entirety. All standards should be met for a student to obtain the optimal level of technological literacy at graduation from high school.
- The benchmarks, which are required for meeting the standards, specify how the student should progress toward technological literacy and what students need to know and be able to do in order to meet the standards.
- The standards must be integrated with one another rather than being presented as separate parts (e.g., Standard 1 with Standard 8 or Standard 19 with Standard 17 and 20).
- *Technology Content Standards* should be included in the curriculum at each grade, both in the technology laboratory-classroom and in other subject areas. Teachers should be familiar with standards preceding and following the grades in which they teach.
- Teachers and curriculum developers should address minority, gender, and equity issues to ensure that students are encouraged and motivated to succeed in the study of technology.
- *Technology Content Standards* should be applied in conjunction with other national, state, provincial, and locally developed standards in technological studies and for other fields of study.
- School systems should begin to move toward a K-12 technology program for all students. (See Appendix D for an articulated curriculum example for grades K-12.)

Administrator's Guidelines for Resources Based on *Technology Content Standards*

A variety of resources, including instructional materials, textbooks, supplies, modules, and kits to aid the laboratory-classroom teacher, are necessary to meet the technology content standards. Administrators should ensure that these resources are age and gender appropriate and progressively more rigorous and richer in content. Each should have a clearly designated level of application specifying one or more of the four grade levels (K-2, 3-5, 6-8, and 9-12).

Resources based on *Technology Content Standards* should:

- allow for innovations and alterations that take into account the changing nature of technology.
- integrate the standards, rather than focusing on a single one.
- provide opportunities for students to make connections among a variety of technologies, thereby helping them develop a common core of technological learning. The resources should also allow for the integration of other fields of study into technological studies.
- incorporate varied methods of assessment in order to provide a broad picture of a student's progress; these assessment measures should allow teachers to compare the progress of different students and to prepare individualized activities that take into account a student's strengths and weaknesses.
- reflect the different standards of the designed world (Chapter 7).
- include various teaching methodologies and student learning styles that address diversity.
- incorporate motivational components that will encourage students to pursue their ideas and complete projects and that ensure lifelong learning for all students.
- include experiences and activities that enhance and promote hands-on learning, including problem-based and design-based learning. These experiences and activities also should be open-ended, requiring students to develop and use technological thinking and challenging them to use and apply it in a variety of settings.
- incorporate knowledge and process activities that enable students to think and do for themselves, as well as to be effective team members.
- include activities that demonstrate to students the need to be adaptable.
- provide opportunities for students to demonstrate their understanding of technology and its value to them and society.

Compendium

A compendium of the technology content standards is presented in Appendix C. The compendium provides a brief summary of the content of technology as presented in the standards and benchmarks by grade-levels.

3 The Nature of Technology

STANDARDS

1 Students will develop an understanding of the characteristics and scope of technology **p. 23**

2 Students will develop an understanding of the core concepts of technology **p. 32**

3 Students will develop an understanding of the relationships among technologies and the connections between technology and other fields of study **p. 44**

3 The Nature of Technology

Everyone recognizes that such things as computers, aircraft, and genetically engineered plants are examples of technology, but for most people the understanding of technology goes no deeper.

Yet because it shapes nearly every part of our lives, a basic understanding of technology is essential to make sense of today's changing world.

Put simply, technology is how humans modify the world around them to meet their needs and wants or to solve practical problems. It can range from building protective shelter and growing food to formulating cancer-fighting drugs and constructing a multi-level network. Technology extends human potential by allowing people to do things they could not otherwise do.

Technological activity is purposeful and directed towards specific goals, and sometimes the results are unintended. The development of a particular technology is influenced by a variety of factors, including the needs of individuals, groups, and society as a whole, as well as by the level of development of related technological components, devices, and systems.

This chapter describes what students should understand about the nature of technology in order to become technologically literate and adaptable. The three standards contained in Chapter 3 address what technology is, its general core concepts, and the relationships among various technologies and between technology and other areas of human endeavor.

STANDARD

Students will develop an understanding of the characteristics and scope of technology.

The word "technology" encompasses many meanings and connotations. It can refer to the products and artifacts of human invention — a videocassette recorder is a technology, as are pesticides. It can denote the body of knowledge necessary to create such products. It can mean the process by which such knowledge is produced and such products are developed. Technology is sometimes used very broadly to connote an entire system of products, knowledge, people, organizations, regulations, and social structures, as in the technology of electric power or the technology of the Internet.

Through their innovations, people have modified the world around them to provide necessities and conveniences. A technologically literate person understands the significance of technology in everyday life and the way in which it shapes the world.

Throughout history, the modification of the natural world has taken on different forms. Understanding these different forms and how the human-made world differs from the natural world can lead to an understanding of human innovation. In the Stone Age, for instance, a major technology involved chipping flakes from pieces of flint to shape useful tools — a task that could be done from start to finish by one person. Today, the products of technology generally involve a much more complicated process, often requiring the efforts of many people to transform an idea or concept to its final, practical form.

Engineers are the professionals who are most closely associated with technology. Although they are the innovators and designers, many other people are involved as well — from distributors, manufacturing and construction workers to operators, managers, regulators, maintenance people and ultimately, consumers.

GRADES
K-2

In grades K-2, students will begin to understand that people use creative or inventive thinking to adapt the natural world to help them meet human needs and wants. Students should be actively engaged in identifying the differences between the natural world and the human-made world, in addition to learning about some of the tools and techniques people use to help them do things. At this grade, students should begin to explore how people have developed ways to shape their world in order to improve comfort, ease workloads, and increase leisure time.

Young children are aware of the world in which they live, but they do not generally know how the technologies they encounter came about. For instance, students may not understand how the food they eat is grown, transported, and processed. By learning how technological developments, such as buildings, highways, telephones, and artificial foods have enhanced the natural world, students can begin to comprehend the vast influence of technology on their lives.

In order to comprehend the scope of technology, students in grades K-2 should learn that

A. **The natural world and human-made world are different.** The natural world includes trees, plants, animals, rivers, oceans, and mountains. The human-made world includes buildings, airplanes, microwave ovens, refrigerators, and televisions.

B. **All people use tools and techniques to help them do things**. By using technology, people adapt the natural world to meet their needs and wants and to solve problems. All people use technology in their jobs and in their daily tasks — from librarians and teachers to truck drivers, homemakers, and police officers.

GRADES

3-5

In these grades, the study of technology should enhance previous learning by increasing the students' understanding of how technology helps people. As students continue to develop a clearer understanding of the natural world as opposed to the human-made world, they will develop an understanding of the differences between technology and science.

When students observe how various things are made, grown, or used, they should begin to see that different processes and techniques are used. Teachers should encourage their students to explore these differences in order to determine those unique qualities. Finding answers to their questions will lead to more questions, which in turn will lead to a deeper understanding of processes and techniques.

In addition, students should investigate how technology has altered people's perceptions of the world. For example, they can explore how television has enabled people to view programs and news releases from any part of the globe, how transportation systems have made it possible to travel across a country in a few hours, and how information technology systems let people search libraries without leaving their desks.

Technological development is shaped by economic and cultural influences. As new technologies appear and some demands are satisfied, the wants of humans change, new ideas and innovations emerge, and the cycle repeats itself. This continuing effort to improve products and systems dictates that technology change constantly, thus leading to both positive and negative implications for people and society. To see this principle in action, students could explore the changing forms of specific products and systems. They might trace, for example, the progression of recorded music from cylinders through records, eight-track tapes, cassettes, compact disks, and laser disks. In this way, they could develop an understanding of how creative thinking and problem solving were used to create new and different ways of recording music to fit the changing technological capabilities.

In order to comprehend the scope of technology, students in grades 3-5 should learn that

C. **Things that are found in nature differ from things that are human-made in how they are produced and used.** For example, the essentials for natural plant growth are sunshine (photosynthesis), air, water, and nutrients, while human-made items require an idea, resources (e.g., time, money, materials, and machines), and techniques. Things found in nature, such as trees, birds, and wildflowers require no human intervention. On the other hand, creating a human-made object, such as a garment, requires human participation and innovation. For instance, the fibers from the bolls of a cotton plant are transformed into cloth through spinning and weaving so that they can be made into a cotton garment.

D. **Tools, materials, and skills are used to make things and carry out tasks.** People make tools to help themselves or others do their work: a cook uses knives to cut vegetables; a gardener uses a hoe to remove weeds; an accountant uses a computer to store information. People also use materials, such as paper, wood, cloth, and stone to make things they use every day. Most people develop the

GRADES
3-5

ability to do common tasks, such as cutting paper with scissors, and some people develop special abilities, like flying an airplane.

E. **Creative thinking and economic and cultural influences shape technological development.** For example, the interests, desires, and economy of a group of people will cause a transportation system to develop in one way and not another. A transportation system for a large city may rely on mass transit, while one in a town might require reliance on personal vehicles, such as bicycles or cars.

GRADES

6-8

Students in the middle-level grades will explore in greater detail the scope of technology. From personal and classroom experience, students will be familiar with specific ways in which technology is dynamic, and teachers should build on this experience by reinforcing the idea that technology is constantly changing.

Classroom activities in grades 6-8 should help students understand that technology enables people to improve current technologies, to further their understanding of other technological ideas, and to develop new technologies. For example, computers are used to develop models before a product is actually made.

In addition, students will learn how creativity is central to the development of products and systems. The development of an invention or innovation is closely related to addressing a need or want. In recent years, however, the development of something new has sometimes preceded the need or identification of a problem. This practice leads to a different growth in knowledge that focuses on the development of the product or system instead of meeting the need or desire of a person.

In order for new technologies to be developed, new knowledge and processes must be developed first. This is often done through research and development (R&D), the practical application of scientific and engineering knowledge for discovering new knowledge about products, processes, and services, and then applying that knowledge to create new and improved products, processes, and services that fill market needs. For example, new knowledge developed about microprocessors by engineers and scientists led to the development of modern computer systems. Companies spend considerable resources on developing new understandings of how things work in hopes of creating new products and systems or improving existing ones. Students will evaluate the commercial application of technology and how economic, political, and environmental concerns have influenced its development.

In order to comprehend the scope of technology, students in grades 6-8 should learn that

F. **New products and systems can be developed to solve problems or to help do things that could not be done without the help of technology.** For example, engines increase the speed at which people can travel, and pumps move water to locations where it is needed. The use of technology sometimes helps to improve personal lives by lessening threats, such as disease, toil, or ignorance. However, the desire or need for a new product or system can cause negative consequences, such as when people travel long hours to work in order to pay for improvements for their homes or child and healthcare.

G. **The development of technology is a human activity and is the result of individual or collective needs and the ability to be creative.** Making life easier involves generating new products and systems through creativity and innovation. For example, from the time of the first gas cook stove in 1936

GRADES
6-8

3

to the time of the microwave oven in 1967, the focus was on simplifying the process of cooking and reducing the time of food preparation.

H. **Technology is closely linked to creativity, which has resulted in innovation**. Most inventions are inspired by perceived needs and wants — the hairbrush, for example. Other inventions are linked to developing creative ideas and the way a person uses them, not necessarily their intended use. For example, the invention of the tea bag grew out of a packaging strategy to replace expensive tin containers. Although tea was packaged in small silk bags to give away as samples, some users thought it was a new way to brew the tea, and thus the tea bag was born. An invention can always be improved, and trying new ideas is often key to that improvement.

I. **Corporations can often create demand for a product by bringing it onto the market and advertising it.** Although market demand generally determines the success or failure of a technology, companies often develop products or systems before a need is identified. In order for a technology to be profitable, there must be a market for it — either preexisting or created through an advertising campaign. The promotion of a product or system often determines its popularity and demand.

VIGNETTE

Creating a Safer Working Environment

The following example uses an unforeseen problem as an opportunity to encourage students to develop and use their understanding of how technology can be used to solve problems. Students in Ms. K's technology class noticed a problem while gluing some wings on their rocket launch system — the smell of the glue. Concerned about the effects of breathing the fumes, the students decided to solve their problem by creating a safer gluing station. [This example highlights some elements of Grades 6-8 *Technology Content Standards* 1, 3, 9, 11, and 20.]

Ms. K first asked the students to define what their problems were and then to brainstorm some possible solutions. The students began listing their concerns:

1. Awful smell
2. Hard to breathe
3. No air circulating
4. Cold outside

After allowing students time to consider their problems, Ms. K divided the class into teams and asked them to identify the most important and pressing concern that could be addressed by designing and building a safer gluing station. After several minutes of working in teams, each group presented its concern to the rest of the class. Each concern was recorded, and the class narrowed them to "awful smell" and "no air circulating."

Next, the students spent two class periods determining the physical limitations of the room and measuring the space necessary for a gluing station. Everyone agreed that the gluing station should be well ventilated and have a safe gluing surface. Students then measured the height of the windows through which the fumes were to be vented, and made rough sketches of the stations using a Computer Aided Design system to refine them.

Once the plans were finalized, the students made a list of the resources they would need to turn their designs into reality: a 3' cardboard box, two 4' x 3' sections of heavy paper, a window fan, a 2' square piece of high-density press board, a roll of duct tape, an electrical cord, a 3' square section of wire mesh, and a 110-volt outlet. The teacher was able to provide recycled and surplus supplies including a discarded double-motored window fan with a bad cord.

After collecting the necessary drawings, materials, and tools, the students created the gluing station by applying skills that they had learned earlier in the year: using form cutting to turn a box into an encasement for the station, attaching the fans to the box using pop rivets, and soldering a cord between the fan and switch. They used a reducer to connect the square hole coming from the fan in the box to a round pipe made from the heavy paper, a device that served as the ductwork leading to the window. Duct tape was used to connect the paper pipe to the cardboard box, and wire was cut to fit the inside of the gluing station to protect fingers from the fan blades.

The students also put their mathematical skills to work to determine if the cubic feet per minute (CFM) rating of the fan was sufficient for the job. After completing the calculations, they decided that the fan did evacuate the fumes quickly enough to avoid inhalation.

GRADES

9-12

In grades 9-12, students will gain a broader perspective of the importance of human innovation and ingenuity in refining existing technologies and developing new ones. They will also continue to develop higher-order thinking skills, such as questioning, investigating, and researching. By the time they graduate, students should have developed an understanding of the scope of technology. This realization includes knowing what technology is and recognizing that it has an intellectual domain and a content base of its own.

Technology is intricately woven into the fabric of human activity and is influenced by human capabilities, cultural values, public policies, and environmental constraints. Students need to recognize these influences and understand how their integration affects technological development. For example, the development of earmuffs was a direct result of harsh, cold winters. Chester Greenwood, a young boy whose ears seemed to be especially sensitive to the cold, decided to develop something new. He designed a special device made of loops of wire and covered with black velvet and beaver fur. Neighbors and friends were so pleased with Chester's invention that they, too, wanted earmuffs, and thus a demand was created, which lead to an 1872 patent for the apparatus. The particular environment, in addition to human activity and capability and the resulting demand, determined the success of earmuffs.

New technologies change people's lives and the way they do things in both expected and unexpected ways. Technological advances build on prior developments and lead to additional opportunities, challenges, and advances in an accelerating spiral of complexity. These advances make modern society vastly different from what was known 10 or 20 generations ago.

Students should realize that inventions occur both by design (e.g., putting a human on the moon) and by serendipity (e.g., 3M Post-it® notes and spin offs). In addition, they should realize innovations are planned and aimed at, such as Edison's light bulb, while others grow unexpectedly out of lines of work that take off in new directions almost as if they have a life of their own. The purposeful application of scientific and technological knowledge speeds up development, while various changes in the physical, political, or cultural environment can act to either speed up or slow down technological development. For example, the appearance of AIDS has spurred research for new vaccines, and the Cold War accelerated the development of both military and space technologies. Ethical concerns have at times restrained the development of certain reproductive and genetic engineering technologies.

Finally, students should understand that the scope of technology involves its essence, its relation to the natural world, and its rapid and often unexpected development. At the same time, students should also understand that the scope of technology includes the commercialization of products and systems. This commercialization frequently results in the development of many inventions and innovations based on market research — who the customers are, what they will purchase, how they will purchase it, and where they will purchase it. The product or system is then prominently presented through advertising to encourage people to purchase it. The intention of advertising is to influence a purchase so that a demand will develop from a desire or an unknown need. The development and marketing of many entertainment devices illustrate such an approach.

In order to comprehend the scope of technology, students in grades 9-12 should learn that

J. **The nature and development of technological knowledge and processes are functions of the setting.** For example, the tractor, plow, and hay bailer are designed specifically for use around farms, while the pick-up truck, tanker, and tractor-trailer are vehicles commonly used to move goods from farms to other areas.

K. **The rate of technological development and diffusion is increasing rapidly.** The rate of development of inventions and innovations is affected by many factors, such as time and money. New technologies are built on previous technologies, often resulting in quick development and dispersion. For example, the first hand-held electronic calculator was designed to perform simple arithmetic. It has quickly evolved from a bulky product owned by a few people to a miniature, multi-function version owned by many.

L. **Inventions and innovations are the results of specific, goal-directed research.** For example, years of research led to the design and development of a laser system used in atmospheric studies. This same laser system was then modified and reapplied to treat the buildup of plaque in the arteries through laser angioplasty.

M. **Most development of technologies these days is driven by the profit motive and the market.** The success of a technology is often determined by whether or not it is affordable and whether or not it works. People often develop and apply technology in a centralized and large-scale fashion to optimize efficiency and reliability, thus resulting in lower production costs.

STANDARD

2

Students will develop an understanding of the core concepts of technology.

Like any other branch of knowledge, technology has a number of core concepts that characterize it and set it apart from other fields of study. These concepts serve as cornerstones for the study of technology. They help unify this study, which could otherwise appear as a collection of ideas that seem only minimally connected, and they provide students with guidance to help them understand the designed world.

The core concepts of technology highlighted by *Technology Content Standards* are systems, resources, requirements, optimization and trade-offs, processes, and controls. Because these concepts are integral areas of technology, they should not be taught as separate topics, but rather, they should be integrated into classes at every opportunity and presented in whatever context is being studied at the time. To that end, the concepts described will also be found interspersed throughout the other standards. In particular, Chapter 7, "The Designed World," will show these core concepts in action in various types of technologies.

The core concepts of technology include:

• **Systems.** A system is a group of interrelated components designed collectively to achieve a desired goal. Systems thinking involves understanding how a whole is expressed in terms of its parts, and conversely, how the parts relate to each other and to the whole. Troubleshooting a malfunctioning system demands considering the various parts and how those parts affect the entire system. Systems should be studied in different contexts, including the design, troubleshooting, and operation of systems both simple and complex.

• **Resources.** All technological activities require resources, which are the things needed to get a job done. The basic technological resources are: tools and machines, materials, information, energy, capital, time, and people. Tools and machines are those devices designed to extend and enhance human capability. Materials have many different qualities and can be classified as natural (e.g., wood, stone, metal, and clay), synthetic (e.g., glass, concrete, and plastics), and mixed — natural materials modified to improve properties (e.g., leather, plywood, and paper). Information, or the organization of data (facts and figures), is critical to the operation of products and systems. Energy involves the ability to do work, and all technological systems require energy to be converted and applied. Capital is the money and other finances available for the creation and use of technological products and systems. Time, which is allotted to all technological activities, is limited, and therefore, its effective use is critical in technological endeavors. Finally, people are the most important resource for all technological activity.

• **Requirements.** Requirements are the parameters placed on the development of a product or system. Requirements include the safety needs, the physical laws that will limit the development of an idea, the available resources, the cultural norms, and the use of criteria and constraints. Criteria identify the desired elements and features of a product or system, while constraints involve the limitations on a design. In addition, knowing how robustness, or over-design, affects the requirements will also aid in developing an understanding of technology.

• **Optimization and Trade-off.** Optimization is a process or methodology of designing or making a product, process, or system to the point at which it is the most fully functional, effective, or as near perfection as possible. The development of the wheel represents a good example of the application of optimization. The entire process of creating should include optimization — from the initial idea to the final product or system. Trade-off involves a choice or exchange for one quality over another. For example, the decision to favor the best material regardless of weight in order to achieve maximum strength may require a designer to make a trade-off of costs. In order to maintain established requirements, trade-offs are made in order to meet the characteristics of an optimum design.

• **Processes.** A process is a systematic sequence of actions used to combine resources to produce an output. An understanding of processes requires time and may not transfer well to other situations without a variety of opportunities in which connections can be made. Designing is the process of applying creative skills in the development of an invention or innovation. The process of making models, as well as modeling in virtual environments, is used to demonstrate concepts and to try out visions and ideas. Maintenance is the process of working with the parts of a system or the system as a whole to ensure proper functioning and to prevent unnecessary errors. Management, which is the process of planning, organizing, and controlling technology, is used to control resources and to ensure that technological processes operate effectively and efficiently. Assessment of products and systems requires asking questions and looking beyond isolated events to deeper patterns. The end goal of assessment is to improve the product or system.

• **Controls.** Controls are the mechanisms or activities that use information to cause systems to change. The household thermostat is an example of a control used to regulate room temperatures. Controls do not always succeed or work perfectly. Understanding the role of feedback, or the use of information about the output of a system to regulate the inputs to a system, is important in being able to determine how controls work in various kinds of systems, such as social, civil, or technical.

GRADES

K-2

Students in the early elementary grades will acquire a basic understanding of many core concepts of technology that will help them as they learn more about the subject. Repeated exposure to those concepts will enable them to make connections and begin to recognize patterns in technological development. They will begin to notice, for example, how the use of a system, such as a heating system, depends on resources and requirements.

Through hands-on activities students will learn that technological activity requires tools, materials, movement, safety, and planning. In addition, they will discover how many of these same concepts relate to other parts of their lives. Students should have as many opportunities as possible to talk about how the technological items they encounter on a daily basis fit into the world around them and how they relate to the core concepts of technology.

The laboratory-classroom should have available a variety of opportunities for students to discuss, explore, and encounter these concepts. Such activities can help students begin to recognize the core concepts of technology.

In order to comprehend the core concepts of technology, students in grades K-2 should learn that

A. **Some systems are found in nature, and some are made by humans.** The solar system in space and the circulatory system in the body are examples of natural systems. One example of a technological system is the information and communication system, which consists of such things as telephones, televisions, printed materials, e-mail on the computer, and letters.

B. **Systems have parts or components that work together to accomplish a goal.** For example, a bicycle can be thought of as a system. It has many parts — wheels, handlebars, pedals, brakes, gears, and chains — and each is important for the bike to function properly.

C. **Tools are simple objects that help humans complete tasks.** Many tools have specific functions, and selecting the right tool makes the task easier. People can use tools to make things. There are many different kinds of tools used in preparing food, such as pots and pans, utensils, and appliances. Children use scissors to cut paper, glue sticks to fasten components together, markers to sketch ideas, and computers to search for information.

D. **Different materials are used in making things.** Paper, wood, cloth, and cardboard are the most common materials children use in making the tools and things they design.

E. **People plan in order to get things done.** For example, children learn that if they want to accomplish something, such as make a birthday card for a parent, they must have the materials available, and they must have the card ready by a given date.

3

GRADES

3-5

In grades 3-5, a strong emphasis will be placed on the concepts of systems, resources, requirements and processes. At this level, becoming more familiar with the core concepts of technology will help students in their development of a total picture of the study of technology. For example, students should be able to identify available resources in their own communities. These resources could include tools and machines in their homes, materials used in building the roads and sidewalks they use when going to and from school, or the information needed to use a new product.

In their mathematics and science lessons, students sort and classify figures, shapes, plants, and animals. Students also should have an opportunity to classify technological systems in order to explore them more easily. Problem solving is another major aspect of mathematics, science, and technology. Through the study of diverse resources and processes, students can develop skills appropriate for solving technological problems. As students get older, they can use more advanced tools to extend their potential. Whether they are using glue guns or computers with design-oriented software, they need to recognize the importance of tools in getting things done.

Introducing the concept of requirements will provide a foundation that will enable students to understand the more complex ideas of later grades. Students will begin to understand the parameters that determine a design or how a product will be developed and used — the safety needs, the physical laws that will limit the development of an idea, the resources available, and the cultural norms. Future discussions of requirements will be related to the use of resources and other core concepts of technology.

In order to recognize the core concepts of technology, students in grades 3-5 should learn that

F. **A subsystem is a system that operates as a part of another system.** An example of a subsystem is the collection of water pipes in a house, which is part of a larger fresh-water distribution system in a town.

G. **When parts of a system are missing, it may not work as planned.** A radio does not work when electricity fails or when a battery has been removed.

H. **Resources are the things needed to get a job done, such as tools and machines, materials, information, energy, people, capital, and time.** Energy transformed into power is used for all technological activities. For example, a battery provides energy to power a flashlight bulb.

I. **Tools are used to design, make, use, and assess technology.** There are many kinds of tools that are used when designing, such as paper, pencils, and rulers or programs specially developed for computers.

J. **Materials have many different properties.** For example wood, stone, metal, glass, and concrete are hard; leather, paper, and some metals can be bent; and glass and some plastics are transparent. The properties of a specific material determine whether it is suitable for a given application.

K. **Tools and machines extend human capabilities, such as holding, lifting, carrying, fastening, separating, and computing.** The use of tools and machines, such as shears, clamps,

GRADES
3-5

vises, carts, drills, saws, and computers makes it possible for people to accomplish more tasks.

L. **Requirements are the limits to designing or making a product or system.** For example, it is often impossible to make a product in a certain way because of the costs of materials or because of time constraints, such as needing the product to be made more quickly than is possible with the method in question. These limits are considered in making decisions about designing and making a product.

VIGNETTE The Bicycle as a Vehicle for Learning

This example uses the bicycle to help students develop an understanding of the various aspects of systems thinking. Students explore the various components of a bicycle and begin to discover how they work together. The teacher leads the students in making connections between their discussion and prior classes. The students explore and work with gears, sprockets, and chains and are given opportunities to tie together conceptual ideas learned in technology and science classes. [This example highlights elements of Grades 3-5 *Technology Content Standards* 2, 3, 5, and 18.]

Mrs. N's fourth-grade class is exploring how various parts of a system work together. She chose the bicycle as an appropriate system because children are familiar with it, and her students had recently learned in a social studies and technology unit that people all over the world rely on bicycles for their transportation needs. The students also had learned in their technology lab that other modes of transportation consume scarce resources and pollute the environment. Mrs. N brought her bicycle in and asked the class to explain how it worked. Bob volunteered that you just get on the bicycle and ride.

Caitlin says, "Well, you are pedaling, and that makes the wheels go."

"But if you don't steer, you'll crash or tip over," Alice adds.

Pointing to the bicycle, Mrs. N restates the children's comments by observing that there are really several things happening: the rider is pedaling and steering, the wheels are turning, and the bicycle is moving.

Mrs. N then uses this explanation to introduce how many systems operate. There is input — in this case, energy coming from the rider pedaling the bicycle; a process, or something happening — pedals making the wheels turn; output — the bicycle moving; and a feedback loop — the rider observing that the bicycle is moving in the direction and speed that is desired.

After discussing the systems and experimenting with the bicycle, Jerome suggests, "Let's call this pedaling system the power system of the bicycle."

"Great idea. Is anything else happening?" Mrs. N asks.

"Well, you have to steer," Jamie says.

"Would you consider steering to be a system also?" Mrs. N asks. Alice thinks for a moment and then answers in the affirmative. "What else?" she asks.

"You have to brake," Jamie notes.

"Would you consider the brakes to be a system also?" Mrs. N asks. After contemplating the question, Jamie decides that the input is the brake lever, the process is the brake pads acting on the wheels, and the output is slowing down or stopping.

"It would appear, then, that a technological system might be made up of several subsystems," Mrs. N concludes. "In the case of the bicycle, there is a power system, steering system, and braking system."

"Thinking about the unit we studied last week on machines, do you observe any simple machines in the bicycle?" Mrs. N asks.

"It has wheels and axles," Megan says.

"The chain sprocket is like a wheel too, and there is a pulley and a lever," Kendall says.

Tomorrow, Mrs. N will help the children discover the relationship between speed and the forces that produce rotation as they experiment with pedaling and shifting the upside-down bicycle. Although the students aren't ready yet for math calculations, Mrs. N plans to have them develop simple charts to depict the relationships.

Gears, sprockets, and chains will also tie the technology and science units together. Some of the students had already experimented with gears and chains by using the constructive building sets during their open discovery time.

GRADES
6-8

After developing a general understanding of the core concepts of technology in the prior grades, students now can investigate these topics and their interrelationships in greater depth. Many aspects of the development and use of technology deal with these systems, resources, requirements, optimization and trade-offs, processes, and controls. Understanding these main ideas will provide a strong foundation for concept development, application, and transfer of technological knowledge in later years.

Students should continue to explore and learn more details about systems, such as the fact that they can take many forms and that systems may have numerous subsystems. Students might investigate, for example, how automated production lines function as a subsystem of the manufacturing system. Simple and complex systems are a vital part of students' lives. Just as the students have organs that will not function apart from their bodies, the parts and subsystems of a technological system will not work properly unless the system is complete. If, for example, a system controlling traffic lights were to suddenly malfunction and cause traffic lights to get out of sequence, the result could be a major traffic jam and many irate citizens.

Most people find it easier to understand how technology works if they see it as a system comprised of connected parts. A new core idea for students at this grade level is systems thinking. Systems thinking is a practice that focuses on the analysis and design of the whole system as distinct from its many parts. Students should learn to look at a problem in its entirety by taking into account all possible requirements and trade-offs. Prior to this level, students have tended to concentrate on the parts that make up the whole. This shift in focus can be difficult, requiring many opportunities for students to develop understanding. Teachers should approach this technique as an introduction to future work in higher grade levels.

Experiences working with different types of technologies and processes help students learn how devices work, as well as how to fix them when they break. This information is used in determining the cause of a malfunction, maintaining products and systems, and managing various aspects of technological development. Understanding various processes requires knowing the context in which a particular process should be used and when it is needed. Therefore, students should have varied opportunities to use many resources, requirements, and processes in order to experience how trade-offs and feedback systems affect results. Students need to learn how to determine if a product, service, or system conforms to specifications and tolerances required by a design.

In order to recognize the core concepts of technology, students in grades 6-8 should learn that

M. Technological systems include input, processes, output, and, at times, feedback. The input consists of the resources that flow into a technological system. The process is the systematic sequence of actions that combines resources to produce an output — encoding, reproducing, designing, or propagating, for example. The output is the end result, which can have either a positive or negative impact. Feedback is information used to monitor or control a system. A system often includes a component that permits revising or refining the system when the feedback information suggests such action. For example, the fuel level indicator of a car is a feedback system that lets the user know when the system needs additional fuel.

N. **Systems thinking involves considering how every part relates to others.** Systems are used in a number of ways in technology. Systems also appear in many aspects of daily life, such as solar systems, political systems, civil systems, and technological systems. Analyzing a system is done in terms of its individual parts or in terms of the whole system and how it interacts with or relates to other systems. For example, discussing a computer system may involve the particular parts of a single computer, or it may include the entire computer network. In contrast, discussing the solar system may involve listing the planets, stars, and other celestial bodies, or it may be discussed by comparing our solar system to other solar systems in the universe.

O. **An open-loop system has no feedback path and requires human intervention, while a closed-loop system uses feedback.** An example of an open-loop system is a microwave oven that requires a person to determine if the food has been heated to the required temperature. An example of a closed-loop system is the heating system in a home, which has a thermostat to provide feedback when it needs to be turned on and off.

P. **Technological systems can be connected to one another.** Systems can be connected with the output of one system being the input to the next system. Sometimes the connection provides control of one system over another system.

Q. **Malfunctions of any part of a system may affect the function and quality of the system.** When part of a system breaks or functions improperly, the results can range from a nuisance to a disaster.

R. **Requirements are the parameters placed on the development of a product or system.** These parameters are often referred to as criteria or constraints.

S. **Trade-off is a decision process recognizing the need for careful compromises among competing factors.** For example, a comparison may be made between increasing the takeoff power of a spacecraft and using lightweight materials. The increased power may result in larger engines, which may be heavier, while the use of the newly developed materials may offset weight concerns. When trade-offs are made, there is a choice or exchange for one quality or thing in favor of another.

T. **Different technologies involve different sets of processes.** For example, data processing includes designing, summarizing, storing, retrieving, reproducing, evaluating, and communicating, while the processes of construction include designing, developing, evaluating,

GRADES
6-8

making and producing, marketing, and managing.

U. **Maintenance is the process of inspecting and servicing a product or system on a regular basis in order for it to continue functioning properly, to extend its life, or to upgrade its capability.** All technological systems will eventually fail. Maintenance reduces the possibility of failing earlier. If maintenance is not done, failure is certain. The rate of failure depends on such factors as how complicated the system is, what kinds of conditions it must operate in, and how well it was originally built.

V. **Controls are mechanisms or particular steps that people perform using information about the system that causes systems to change.** The essence of a control mechanism is comparing information about what is happening to what is desired and then adjusting devices or systems to make the desired outcomes more likely. For example, a microprocessor may be used to control the performance of a microwave or traditional oven in cooking food to a desired temperature.

GRADES
9-12

By the time students enter high school, they should be familiar with the core concepts of technology. At this level they can begin to analyze how those concepts interact in issues that affect them, their community, and the world. Such cross-theme topics as how resources can be sustained and how resources are related to requirements or optimization considerations should be discussed and explored in great detail.

Students should focus on the concepts of systems analysis, stability of systems, and control systems. They should recognize that the order in which processes are used is variable and that new technologies are often created out of existing ones. The marketing of these new technologies has a direct effect on future developments and innovations.

Students need to shift from focusing on how the development of technology affects them locally to a broader, global outlook. The use of systems thinking requires students to examine all aspects of a problem, such as its criteria, constraints, benefits, and consequences. Using systems thinking helps students to determine if the development of a particular system is worth the efforts and cost, and to determine the best approach. Resources can also be examined from a global perspective by exploring the sustainability of the Earth's resources. The management of work and resources is a major factor in the success of the commercial applications of products and systems. Poor management can lead to excessive costs, poor quality, and inefficiency. Good management helps ensure that processes and resources operate effectively and efficiently. The use of schedules, in addition to the allocation of material and space, affects the use of many technologies.

Students should learn that the processes of technology do not always happen in a linear order. For example, prototypes, which are often made as part of the design process, are used to help assess the quality of a design before the product or system is actually made and used. Likewise, students need to understand that new innovations are seldom ready for market. Once innovations have been designed, they must be tested and prepared for future use. Because of requirements — capital, timing, demand, and production problems, for example — not all technologies make it to market. The lifecycle of a product (or system) includes the processes from concept to eventual withdrawal from the marketplace. Some product lifecycles are quite long while others may be very short.

Optimization and trade-offs are topics that require more time and effort for students to develop an understanding of their importance in technological development. Students should have opportunities to use simulation or mathematical modeling, both of which are critical to the success of developing an optimum design. If a mathematical model is not possible, then students will have to rely on their personal experience and use of physical models. Students will need to recognize the limitations of physical models and the limits their use imposes on being able to make various adjustments. Likewise, students need repeated exposure in determining trade-offs because this important principle will be encountered in many areas of science, in addition to technology.

GRADES
9-12

Finally, the study of controls involves simple as well as complex systems. The human body includes controls that determine breathing, circulation, and digestion. These systems in nature are much more complicated and sophisticated than the most advanced human-made control systems. Reliability, feedback, and the basic function of a control device determine how efficient and beneficial it proves to be. Therefore, students need exposure to an array of experiences and activities which focus on designing and working with control systems.

In order to recognize the core concepts of technology, students in grades 9-12 should learn that

W. **Systems thinking applies logic and creativity with appropriate compromises in complex real-life problems.** It uses simulation and mathematical modeling to identify conflicting considerations before the entire system is developed.

X. **Systems, which are the building blocks of technology, are embedded within larger technological, social, and environmental systems.** For example, a food processor is a system made up of components and subsystems. At the same time, a food processor is only one component in a larger food preparation system that, in turn, is a component in a larger home system.

Y. **The stability of a technological system is influenced by all of the components in the system, especially those in the feedback loop.** Cruise control in an automobile, for example, automatically detects and controls the speed of the car. Some delay in feedback or in functioning can cause a cycle to develop in a system.

Z. **Selecting resources involves trade-offs between competing values, such as availability, cost, desirability, and waste.** Technological development involves decisions about which resources can and should be used. For example, some homes are very energy efficient, while others consume large amounts of energy.

AA. **Requirements involve the identification of the criteria and constraints of a product or system and the determination of how they affect the final design and development.** Sometimes requirements can be constraints, criteria, or both. Balancing the two is the optimum.

BB. **Optimization is an ongoing process or methodology of designing or making a product and is dependent on criteria and constraints.** Optimization is used for a specific design purpose to enhance or to make small gains in desirable characteristics. An optimum design is most possible when a mathematical model can be developed so that variations may be tested.

CC. **New technologies create new processes.** The development of the computer has led to many new processes, such as the development of silicon chips, which led to smaller-sized components.

DD. **Quality control is a planned process to ensure that a product, service, or system meets established criteria.** It is concerned with how well a product, service, or system conforms to specifications and tolerances required by the design. For example, a set of rigorous international standards (ISO 9000) has been established to help companies systematically increase the quality of their products and operations.

EE. **Management is the process of planning, organizing, and controlling work**. Management is sometimes called getting work done through other people. Teamwork, responsibility, and interpersonal dynamics play a significant role in the development and production of technological products.

FF. **Complex systems have many layers of controls and feedback loops to provide information.** Controls do not always succeed or work perfectly. The more parts and connections in a system, the more likely it is that something may not work properly; therefore, human intervention may be necessary at some point.

STANDARD

3

Students will develop an understanding of the relationships among technologies and the connections between technology and other fields of study.

The products of technology are used in every field of study. Technological progress often sparks advances and sometimes can even create a whole new field of study. For example, the telescope made possible the era of modern astronomy, and the movie camera led to a whole new art form. Conversely, technology borrows from and is influenced by many other areas. There may be no field of study as intimately connected with so many other fields as technology.

Technology has its own unique content base with specific concepts and principles that set it apart from these other fields. Technologies are intimately related, such as the manufacturing used to produce generators and motors that are then used in energy and power technology. Because technology cannot really be appreciated in isolation, students need to understand that these interrelationships exist and to gain an appreciation for how the relationships shape technology.

Standard 3 discusses various opportunities to connect ideas and procedures that demonstrate how technologies are interrelated and combined. This standard also addresses how new products and systems build on previous inventions and innovations, while demonstrating how knowledge acquired in one setting can be applied in another. For example, understanding how to mass-produce a biological product developed in a research laboratory is essential to the building of a biotechnology company. The biotechnology industry has learned that there is a vast difference in engineering a product in a laboratory and mass-producing it for customers. Research about the various efforts addressing production problems associated with bioprocesses is proving to be vital.

Science and technology are like Siamese twins. While they have separate identities, they must remain inextricably connected in order to survive. Science provides the knowledge about the natural world that underlies most technological products today. In return, technology provides science with the tools needed to explore the world. The two fields have many similarities, such as the development of codified sets of rules and reliance upon testing of theories in science and of designs in technology. The fundamental difference between them is that science seeks to understand a universe that already exists, while technology is creating a universe that has existed only in the minds of inventors.

Mathematics and technology have a similar but more distant relationship. Mathematics offers a language with which to express relationships in science and technology and provides useful analytical tools for scientists and engineers. Technological innovations, such as the computer, can stimulate progress in mathematics, while mathematical inventions, such as numerical analysis theories, can lead to improved technologies.

Other fields of study also have relationships with technology. The designers of bridges, dams, and buildings are often influenced by art forms. In turn, technology affects the humanities, often quite profoundly, with inventions that offer new capabilities and approaches. For example, the synthesizer and the computer have aided in the composition and performance of music, while computer databases have revolutionized research in the social sciences.

GRADES
K-2

Learning becomes more meaningful when students can connect knowledge gained in the classroom to their everyday experiences. The study of technology provides many opportunities to make such connections. As students establish these connections early in their education, they will begin to understand how technology influences their daily life.

Because the study of technology has numerous relationships with other areas of the K-12 curriculum, it is particularly important to introduce technology at this level. Teachers can focus on the common ground between technology and other subjects (e.g., science, mathematics, social studies, language arts, health, physical education, music, and art).

One effective method is to use themes from familiar literature, such as *Richard Scarry's How Things Work, Charlotte's Web, The Three Little Pigs*, or *Jack and the Beanstalk*, to make connections with the study of technology. For example, E.B. White's novel, *Charlotte's Web*, could offer an opportunity to learn about such connections.

Once the book is read in class, students could use photographs, drawings, or actual spider webs to examine and describe the design of various webs. They could copy a particular design using materials, such as yarn, string, or strips of construction paper and then decide which materials were easiest to use, best suited for the design, or provided the best results. Classroom discussions about the novel could provide opportunities for students to build connections between science and the study of technology: How do spiders make webs? How do they use their webs? Why are the small strands in webs so strong? How do humans apply similar designs (nets, for example)? The children's classic, *The Three Little Pigs*, could provide the inspiration for students to build models of each house and then test them for strength and durability. By understanding which structure is the best for the pigs, even very young students should grasp concepts such as the properties of materials, construction techniques, measurement, and scale.

Through activities in grades K-2, students will have the opportunity to explore, discover, and make connections between technological studies and other fields of study — an important component in the process of learning and understanding about the value of technology to society and culture. Through the combined investigation of these fields of study, students will develop a well-rounded knowledge base.

In order to appreciate the relationships among technologies, as well as with other fields of study, students in grades K-2 should learn that

A. **The study of technology uses many of the same ideas and skills as other subjects.** For example, many ideas learned in mathematics are also used in the study of technology, such as basic rules of numbers and using numbers to represent measurements. The use of ideas or skills learned in the study of technology, such as measuring and building an object, may be used to build a representation of data collected during mathematics instruction.

VIGNETTE

Helping Out Stuart Little

This example uses the story line of *Stuart Little*, by E. B. White, to help students develop an understanding of how the study of technology relates to other fields of study and vice versa. The students develop a basic understanding of how things work and how the topics they learn in school are related. [This example highlights some elements of Grades K-2 *Technology Content Standards* 1, 2, 3, 10, 11, and 20.]

To provide continuity throughout the school year, Ms. L, a second-grade teacher, chose to use *Stuart Little* by E. B. White as the basis for many of her classroom activities. By using this popular children's story, she was able to introduce her students to many concepts in the study of technology, as well as other subject areas.

While she was reading the book to the students, they decided the Little family needed a new home, one that would fit Stuart and his family. Ms. L used this opportunity to discuss with the class the differences between fundamental needs (shelter, food, and clothing) and wants.

Working in teams, the students selected different rooms that they would build for the Little home. Each team selected a cardboard box that was the appropriate size and shape for the room that they were developing. Using the materials provided, the students constructed doors, windows, stairs, and furniture for the Littles. Next they chose colors and materials to decorate their particular room, and with the aid of their teacher, they used simple electrical devices to wire their model house for lights.

When they read about the Big Cat coming around, the students, with the guidance of Ms. L, applied what they had learned about safety and protection to develop a security system. Using aluminum foil, cards, wire, lights, and buzzers, they designed and assembled an early warning cat-alarm system that would alert the Little family when the Big Cat stepped onto the porch.

Later in the year, the students planned a local trip to the zoo for the Little family. They built different vehicles for the family's travel and used maps to determine where the zoo was located, how far they would have to travel, and how long it would take to get there.

In the spring, the students planned a trip to China for the Little family. They located China on a map and determined how far away it was. They also learned about Chinese customs, and the value of U.S. money in China, and even constructed passports so the Littles could get through customs.

GRADES

3-5

As a result of their experiences in grades 3-5, students will have an understanding of and an appreciation for many possible relationships that exist among technologies and between the study of technology and other fields of study. Students then need to develop the confidence to apply these relationships to help them construct a greater understanding of how things work, how the designs of many technologies use natural phenomena, and how technologies affect the development of other technologies.

Various technologies are often combined in the development of new products and machines. Mechanical parts, such as springs, wheels, belts, gears, levers, and cam and cranks, for instance, are used to make simple machines that in turn are combined to produce more complex machines and systems. The combination of technologies is not always obvious. It is hard to see the components that make up a microwave oven or a high-powered microscope. However, the combinations of technologies are more obvious in other devices — a roller coaster, for example.

Although technology is a human enterprise with a content and history of its own, it is interdependent with other fields of study. By creating laboratory-classroom environments where this interdependency is highlighted, teachers can increase the opportunities for ideas to flow naturally from lessons in one subject to lessons in another. For example, rockets and space fascinate many children and offer a natural opportunity for teachers to bring together several fields of study. Students could begin by studying about the moon's surface and movement in their science lessons. Next, they could take a historical look at the development of various rockets. The students could then design a rocket and build a model to test their design. They could apply their estimation skills learned in a prior mathematics lesson to determine how far their rockets could fly. Finally, they could write a creative paper describing what it would be like to be an astronaut traveling in space. By seeing these connections made in the classroom, students would gain a clearer understanding of why they need to learn certain concepts and principles.

In order to appreciate the relationships among technologies, as well as with other fields of study, students in grades 3-5 should learn that

B. **Technologies are often combined.** For example, an escalator uses the wheel and axle, inclined plane, pulley, gears, belts, and an electric motor to move people from one level to another.

C. **Various relationships exist between technology and other fields of study.** For example, the study of technology includes the study of natural scientific laws, systems, design, modeling, trade-offs, and side effects. These topics are also explored and studied in science and mathematics. Likewise, numerous fields of study share the common concepts of communication, scale, constancy, and change.

GRADES
6-8

Through the study of technology, students in grades 6-8 begin to discover the answer to the perennial question, "When am I ever going to use this knowledge?" The study of technology in the middle-level grades helps students recognize relationships among different topics in technology, make connections across fields of study, and integrate ideas and activities in a structured setting.

For example, learning about aviation technology could involve the study of key technological design and performance characteristics of planes and helicopters, mathematical calculations used for altitudes and airwave movement, and technical documentation of the history of aviation and scientific principles of flight. This information could help students make connections across various subjects or disciplines, while providing opportunities for them to expand their knowledge through making, testing, and exploring their designs in the technological studies laboratory-classroom.

Students need various opportunities to explore how technological ideas, processes, products, and systems are interconnected. For example, in the healthcare system technological devices that monitor the heart, blood pressure, and breathing are dependent on other technological devices, software, and hardware in order to perform properly. In the home, heating systems are dependent upon a thermostat system. If one aspect of a system is not functioning properly, the entire system may malfunction or break down. Students also can study relationships within technology by exploring the role of various occupations, such as engineering.

The free sharing of processes and techniques has generally been limited to the federal government. Because of economic interests, businesses and industry do not typically give ideas away. Rather, they rely on patents to protect ideas so that others cannot copy them without permission. The sharing of technological knowledge is viewed as a means to improve the quality of life and bolster the country's competitiveness in the global marketplace. For example, the ideas developed in genetic engineering, first used in plants in space program research, are now being transferred to research and new developments involving human tissue.

Students should be encouraged to look for relationships between the study of technology and other fields of study. They should understand that knowledge gained in one field of study could be applied to another. Such experiences will enable students in grades 6-8 to develop systems thinking by understanding how parts work together to form the whole.

In order to appreciate the relationships among technologies and other fields of study, students in grades 6-8 should learn that

D. **Technological systems often interact with one another.** In automated manufacturing, for example, computer systems interact with manufacturing systems.

E. **A product, system, or environment developed for one setting may be applied to another setting.** For example, a computerized pump based on biological laboratory design for the Mars *Viking* space probe was modified for use as an insulin delivery mechanism that provides diabetics with an automatic and precise way to control blood sugar.

F. **Knowledge gained from other fields of study has a direct effect on the development of technological products and systems.** Studying the history of technology provides people with a way to learn from the successes and failures of their predecessors. In addition, skills learned from other fields of study enhance technological developments. For example, skills learned in language arts are used in making design presentations. The concepts and principles of drawing are used in designing and rendering examples of technological products and systems. Scientific and mathematical knowledge and principles influence the design, production, and operation of technological systems. Science concepts, such as Ohm's Law, aerodynamic principles, and the periodic table of elements, are used in the development of new materials and designs. Mathematical concepts, such as the use of measurement, symbols, estimation, accuracy, and the idea of scaling and proportion are key to developing a product or system and being able to communicate design dimensions and proper function.

GRADES

9-12

At the 9-12 grade level, students will increase their understanding of technology to include an in-depth understanding of its relationships and connections. Developing an appreciation for the vast relationships in technology will help students begin to understand how future developments and society's well-being is dependent on how well technology is understood, developed, used, and restricted.

The sharing of the development and production of an invention (a new product or system) or innovation (an improvement of an existing product or system) expands the knowledge base of technology. This new knowledge base will have a direct effect on the ability of people to develop and produce more technologies, which is referred to as technology transfer. Technology transfer, or spin-offs, is an exciting concept for students. Students need to have various opportunities to investigate how technology transfer happens within a technology, among technologies, and across other fields of study. They also need to study the economic benefits of technology transfer. By using various resources to gather information on technology transfer, for example, students can prepare a presentation that demonstrates how technology is transferable, its potential for new applications, and its benefits.

In the highly competitive world of business, obtaining patents is critical. Patents are designed to protect the financial potential of an idea, invention, or innovation by prohibiting others from copying the processes and final products unless they provide financial compensation to the developer, or unless they wait until a given period of time has passed. In contrast, scientific knowledge is often communicated openly to the public through presentations at professional meetings and publications in scientific journals.

Science, mathematics, engineering, language arts, health-related fields, fine and performing arts, and social studies offer direct connections to technology content. Teachers in these fields of study can include the use of tools, artifacts, resources, simulations, and computer models to better illustrate the knowledge or concepts they are teaching. Likewise, students in a technology laboratory-classroom can use content from other fields of study when studying about technology.

For instance, when students in a physical education class are discussing ergonomics (the study of the body as it relates to design), they could build on experiences by applying ergonomic principles in their technology laboratory-classroom. They could then relate information learned in their physical education class about stresses applied to the body, combined with knowledge from science and mathematics dealing with the forces of motion, to design a model of an amusement park ride or functional furniture. Connecting and synthesizing technological knowledge with other fields of study can provide valuable information for students as they learn more about the world around them.

In order to appreciate the relationships among technologies, as well as with other fields of study, students in grades 9-12 should learn that

G. Technology transfer occurs when a new user applies an existing innovation developed for one

GRADES

9-12

purpose in a different function. Aerospace composite materials, for instance, were used to design an advanced wheelchair that proved to be lightweight and easy to maneuver.

H. **Technological innovation often results when ideas, knowledge, or skills are shared within a technology, among technologies, or across other fields.** The sharing of knowledge about irrigation techniques, for instance, can enable developing countries to try out new ideas and make innovative adjustments to their current systems to improve the delivery of water.

I. **Technological ideas are sometimes protected through the process of patenting.** The protection of a creative idea is central to the sharing of technological knowledge. Most often an idea is protected through the long and tedious process of obtaining a patent. The purpose of a patent is to safeguard the investment of the inventor or creator and to give credit where and when it is due.

J. **Technological progress promotes the advancement of science and mathematics.** Likewise, progress in science and mathematics leads to advances in technology. The development of binary language, a digital language made up solely of ones and zeros; the invention of the transistor, a device designed to replace the vacuum tube; and the use of integrated circuits, a collection of millions of miniature transistors, helped spawn a new generation of machines, from laptop computers and compact disc players to digital television. The mathematical and scientific ideas applied in the development of these digital devices promoted further developments that resulted in new tools, such as computer modeling. These tools, in turn, are used to explore new scientific and mathematical ideas, thereby spawning additional discoveries.

VIGNETTE

A Hands-On Experience

This example demonstrates how collaboration and coordination of curriculum allows students to make connections and understand the relationships of the study of technology with other fields. [This example highlights some elements of Grades 9-12 *Technology Content Standards* 1, 3, 11, 12, and 14.]

The science teacher, Mr. C, and the technology teacher, Ms. M, worked together to develop a unit concerning joints, tendons, muscles, and prosthetic devices.

In their science class, students learned how these different body parts function by using a human skeleton and related pictures to identify each part of the body.

In their technology class, the students learned about the development of prosthetic devices from a historical standpoint. The students then divided into groups to fabricate a prosthesis for one of their own hands. Each group received a poster board, string, elastic strips, straws, glue, and a utility knife. The criteria and constraints stated that the hand must be able to pick up a table tennis ball, pick up a piece of paper lying flat on a table top, dial a rotary phone, and be displayed in a stand-alone presentation. After a couple of class periods, the students tested their "hands" and determined if they met the criteria and constraints.

As a result of the learning experiences in both classes, students developed a clear understanding of the unique functions of joints, tendons, and muscles, in addition to how various technologies were used in the development and operation of prosthetic devices that mimic body functions.

4 Technology and Society

STANDARDS

4 Students will develop an understanding of the cultural, social, economic, and political effects of technology **p. 57**

5 Students will develop an understanding of the effects of technology on the environment **p. 65**

6 Students will develop an understanding of the role of society in the development and use of technology **p. 73**

7 Students will develop an understanding of the influence of technology on history **p. 79**

4 Technology and Society

To be understood properly, technology must be put into a social, cultural, and environmental context.

To a large degree, a society will determine the wants and needs that the use of technology seeks to address. This in turn shapes the paths that technological development will take. The physical environment, too, can play a role by creating constraints or causing certain needs. The initial development of the steam engine, for instance, was driven by a need to pump water out of coal mines, and the coal mines were needed because most of the wood in British forests had already been burned for fuel.

Conversely, technology affects both society and the environment. Technology has been called "the engine of history" for the way in which its use drives changes in society; it influences cultural patterns, political movements, local and global economies, and everyday life. And, as technology has grown to meet the demands of the world's billions of people, its power over the environment has grown as well, to the point where its use has the potential both to improve or to cause great damage to the environment.

In general, the effects of society on technology and technology on society go hand in hand, so that the two march together toward the future. The invention of the personal computer, for example, was driven by the interest of a small number of hobbyists. Once the computer was invented, people in the business world and general population began to find uses for it. This sparked more development, which made it useful for greater numbers of people, whose interest drove further development, and so on, in an ever-accelerating spiral of adoption and development.

This chapter deals with how the use of technology affects society and the environment, as well as how society influences the development of technology, and how technology has changed and evolved over the course of human history.

STANDARD

Students will develop an understanding of the cultural, social, economic, and political effects of technology.

In many ways, technology defines a society or an era. This delineation is reflected in the naming of time periods to reflect the respective technological milieus: Stone Age, Bronze Age, Iron Age, Industrial Age, Information Age, and so forth. Technology shapes the environment in which people live, and over the course of time, it has become an increasingly larger part of people's lives. Meanwhile, the natural components of the environment have become correspondingly smaller. Most people in our country live in houses or apartments, work and shop in large buildings, move about in vehicles, eat prepared foods, drink water from a public system, and rely on newspapers, radio, television, and the Internet for much of their communication. People occupy a technological world.

Many of technology's effects on society are widely regarded as desirable. Advances in medicine and public health have enabled people to live longer, healthier lives while eradicating diseases that once prevented many children from living to adulthood. Public water and sewer systems have provided water to remote areas and removed contaminants. Improved transportation and communication systems have brought the world closer together, and automated manufacturing systems have allowed the average citizen to own cars, televisions, computers, and a host of other consumer goods.

Other effects of technology are sometimes regarded as less desirable. Traditional ways of life have been displaced by technological development. This trend tends to magnify the inequalities among peoples and among societies by creating a situation in which a minority of people and groups control and use a majority of the world's resources. As the pace of technological change continues to increase, questions arise as to whether society's political and social norms can effectively keep up with the changes.

Such factors dealing with the use of technology make it important that decisions be made with care about any particular product or system. For instance, the emerging technology of genetic engineering has great potential for improving agriculture and the treatment of disease, it carries a number of concerns and ethical quandaries as well. In a democratic society such as ours, individual citizens need to be able to make responsible, informed decisions about the development and use of such technologies.

GRADES

K-2

By the time they enter kindergarten, students have already been exposed to various home appliances, building tools, communication products, and transportation vehicles. A natural place for students to start learning about technology is by having them reflect on how they use such things. For example, a teacher might ask students to identify products and systems they use and how they use them (e.g., the refrigerator keeps food cold, the microwave oven cooks food, and the television provides entertainment).

Through guided inquiries, observations, and discussions, students can also become aware of other forms of technology in their lives, how they are used, and what makes them effective. For example, students could explore what bell is used for recess and how that bell is different from the bell used in a fire drill. Building an awareness of how technology is connected to each person's life provides a foundation for a later exploration of its effects.

Students should be encouraged to look at both the positive and negative results of the use of technology. Although products and systems are generally designed to enhance life and improve living conditions, the outcome is not always positive. Laboratory-classroom activities can help students recognize that when products do not work as planned, problems can be created. Likewise, students should look at technologies that enhance life and improve living conditions. For example, teachers could encourage students to ask questions to determine the positive and negative effects of artificial light. The students could examine how home, office, and street lighting is used to illuminate dark environments and provide protection and security. They also could examine how different kinds of light fixtures are used in a variety of situations.

In order to recognize the changes in society caused by the use of technology, students in grades K-2 should learn that

A. **The use of tools and machines can be helpful or harmful.** Scissors can be used to cut paper, but they can also cause injury. A wagon can be used to haul toys, but if the wagon tips, the toys will spill and may break.

GRADES

3-5

Students are eager to know about the world around them — how things work, or why they work the way they do. They ask questions, such as: Why does a plane fly, and how is it built? How did early people measure the length of something? How do escalators and elevators work? What makes a computer work the way it does? Learning how technology influences or changes their lifestyles takes time and experience. Providing opportunities for students to explore, ask questions, and use information resources allows them to begin to find answers to their technology questions, which in turn will lead to more questions and more answers. As they explore and make connections, students begin to build a knowledge foundation they will use later in solving problems and understanding the influence of the use of technology on society.

For instance, students who have been studying pulleys and counterweights might then investigate how an elevator operates. After building a model of an elevator, they could see how pulleys and counterweights work to create a machine that can move people and goods up and down. From there, they could discuss how such machines improve the mobility of people and goods and the effect that elevators have on the design and construction of buildings.

Students can consider the issues surrounding transportation, land use, pollution control, and communication to become knowledgeable about the decisions made during their development. In examining how such decisions are made, students should recognize that the use of technology results in both expected and unexpected consequences. They might, for instance, discuss landfills and how poor design or construction has sometimes led to the contamination of surrounding soil and water. Through these exercises, students will learn that making sound decisions demands examining both the costs and benefits of technological development.

In order to recognize the changes in society caused by the use of technology, students in grades 3-5 should learn that

B. **When using technology, results can be good or bad.** These results may affect the individual, family, community, or economy. An example of a good result is using air conditioning to help keep cool. However, during a heat wave, the overuse of air conditioning can result in a power outage, which can leave a community without electricity. Ships transport oil, which helps people by supplying fuel for homes, cars, and other things. But when a ship wrecks and oil spills into the ocean, the environment can suffer immeasurably.

C. **The use of technology can have unintended consequences.** When a dam is built for the purpose of supplying water for a city, it can also provide a habitat for plants and animals uncommon to the area. At the same time, covering a large area with water can destroy native plants and animals. Developers must decide whether the product or system will be helpful, and if so, what the best plan will be to put it into use.

GRADES

6-8

In the middle-level grades, students will discuss how technology causes cultural, social, economical, and political changes in society, with an emphasis wherever possible on how the use of technology influences their own lives. For example, students could examine how technology used in education has changed their learning environments. They also could reflect on how their safety and comfort are enhanced by new products and systems in buildings and classrooms. Likewise, students could determine how the use of certain technologies affects choices and attitudes of school personnel and the students themselves.

Students should understand that technology itself is neither positive nor negative, but that the use of products and systems can have both desirable and undesirable consequences. When technologies work as intended, the consequences can be desirable, such as providing comfort from the elements, mitigating diseases, and using natural resources more efficiently. Sometimes, however, the consequences are undesirable, such as loss of jobs, the loss of resources, or the misuse of time. Some children, for instance, spend many hours watching television or playing video games instead of doing their homework or exercising. By investigating such issues, students will come to understand the various roles of technology and the value of its use in society. For example, students could be taught how the development of motion pictures led to the creation of the movie industry, which in turn has affected the economy, particularly in southern California.

Understanding the effect that the use of technology has on cultural, social, economical, political, and ethical issues is another important concept. Exploring such issues will provide students with opportunities to consider principle concerns, employ critical questioning, and determine the benefits and changes in society caused by the use of different technologies. Such exploration will enhance their reasoning, logic, and critical thinking skills.

In order to recognize the changes in society caused by the use of technology, students in grades 6-8 should learn that

D. **The use of technology affects humans in various ways, including their safety, comfort, choices, and attitudes about technology's development and use.** People's attitudes toward and knowledge about a product or system, along with their subsequent actions, vary greatly and are influenced by their moral, social, or political beliefs. For example, some might support the construction of a high-voltage electric transmission line because it would provide electricity to people in remote areas, while others who live near the path of the power line might not support it because of potential effects on their health and safety. Sometimes people are well informed about a product or system, while at other times they have limited information to make their choices about whether a technology should be developed or used.

E. **Technology, by itself, is neither good nor bad, but decisions about the use of products and systems can result in desirable or undesirable consequences.** For example, fossil fuels have both desired and undesired consequences. While these fuels provide a good source of energy, their use may damage the environment.

F. **The development and use of technology poses ethical issues.** People often wonder whether the use of some technologies is ethically acceptable. For example, should we allow everyone to purchase a gun?

G. **Economic, political, and cultural issues are influenced by the development and use of technology.** For example, information technology systems have been used to both inform and influence society. Technology also affects the way people of different cultures live, the kind of work they do, and the decisions they have to make.

GRADES

9-12

Students at this level will continue to study how the development of various technologies affects cultures and societies in both subtle and obvious ways. Working from this foundation, students will learn that the changes caused by technology have been driven by the desire to improve life, increase knowledge, and conquer time.

New technologies are often developed in response to an identified need or want or a technological demand. But some novel products and systems, such as some entertainment devices, medicines, and foods, have emerged as a result of the application of new technological knowledge or techniques. Students should explore these emerging technologies and develop the skills to evaluate their impacts. They should learn to reason and make decisions based on asking critical questions, not on the basis of fear or misunderstandings. The goal is to equip them with the necessary knowledge and the proper mental tools to be able to examine technological issues and come to their own conclusions in a responsible, ethical manner.

A classroom activity might, for instance, have students explore the use and development of synthetic rubber and its related products, such as nylon. Teachers could guide students to recognize the various decisions and issues that were a part of the development process and the effects initiated by world events. For example, the study of World War II could provide an opportunity to discover why there was a need for synthetic rubber to replace natural rubber. Due to the war, resources were limited because of military needs, and natural products were not readily available. Thus, experimentation and new developments grew out of urgent wartime needs.

Finally, students need to recognize the value of transferring technological knowledge within and among cultures and societies. They should be able to point out how the transfer of technology from one society to another affects other cultures, societies, economies, and politics.

In order to recognize the changes in society caused by the use of technology, students in grades 9-12 should learn that

- H. **Changes caused by the use of technology can range from gradual to rapid and from subtle to obvious.** Those changes have resulted in people having information overload, rapid adaptation or acceptance of short-lived relationships, and the need for instant gratification. For example, when people listen to a classic album or watch television on their high-tech entertainment system, they are able to program segments of the album to play in a certain sequence or watch two television programs at once while they preview the highlights of a third and record a fourth.
- I. **Making decisions about the use of technology involves weighing the trade-offs between the positive and negative effects.** These decisions can have lasting impacts, sometimes affecting living habits and cultural patterns on a global scale. The construction and use of the interstate system require considering the benefits of providing a safe and quick mode of transportation, as well as the effects on the economy and society.

J. **Ethical considerations are important in the development, selection, and use of technologies**. For example, medical advances for prolonging life and treating illness have triggered concerns about health care providers giving more attention to the best technological solution than to human values or needs. Questions about how medical technologies should be used to sustain life and the related costs must be considered. High-tech medicine has transformed the philosophy of doing everything possible to prolong life into a consideration that living longer may not necessarily mean living better.

K. **The transfer of a technology from one society to another can cause cultural, social, economic, and political changes affecting both societies to varying degrees.** Sharing methods to increase food production and preservation can alter a country's living habits in significant ways. For example, the idea for developing flash freezing, a method to freeze foods that preserves the flavor, appearance, and nutritional value, was based on how the people of Labrador preserved their food. The resulting invention, frozen food that is ready to heat and eat, has considerably changed the living habits and culture of many societies.

VIGNETTE

This example involves students learning about the various issues raised during the development of a local airport site. Students are encouraged to consider all issues and to look at how the use of technology causes cultural, social, economic, and political changes in their own area. They are asked to work in collaboration with students in a biology class and to make a joint presentation of their findings. [This example highlights some elements of Grades 9-12 *Technology Content Standards* 4, 5, 6, 10, 11, and 18.]

Students Plan New Airport Site

Students in Ms. T's technology class were discussing the issues surrounding the development of the new regional airport near their school. The students thought it would be a good idea to design a layout of the airport and see how their plan compared with that of the developers. Ms. T asked students to review aerial photos of a practical site, outline the area on a land plot book, and sketch a geographical map of their proposed site.

The students soon discovered just how much the airport would affect the region. Because of the amount of acreage needed, a state highway would need to be rerouted, part of a creek bottom would need to be rechanneled, and many farms would need to be bought. The students voiced concern about the impact on the local economy, the environment, and political issues, as well as the relocation of residents. After much discussion, the class decided that a good science activity would involve studying about wetlands preservation and the pollution that the airport could bring to the area.

Mr. D's biology class joined in on the project and began looking at the effect of the proposed action on the various species of wetland animals. Additionally, the biology students designed a survey to send to the residents near the site to obtain information concerning how the citizens felt about the proposed airport project. In another activity, students participated in a field trip to a regional airport 50 miles from their site and recorded sounds at varying distances from the airport. Mr. D then asked the class to divide into groups and develop reports on the wetlands, resident life, noise pollution, and political issues.

After the two classes had worked independently for several weeks, the technology and biology students met jointly to share their work. The technology group had completed a CAD layout based on proposed plans for the airport, which included details of the terminal, runways, control tower, and support facilities. One group of students had completed a new land plot map to show the geographical changes the airport would create in the rural area. Another group had created a scale model of the airport, while yet another had designed a rerouting plan for the state highway.

The biology students presented a report on the wetlands supported by photos, graphs, and charts of the animals, which would be endangered by the airport development project. Students who had conducted the survey about the impact of the airport on residents' lifestyles presented another report. Yet another group attended local hearings and prepared a report that outlined the political issues of the proposed airport. The final report contained a probability study of noise pollution on the surrounding area.

Because the information developed by the two classes was so significant, the teachers encouraged the students to combine their findings into a comprehensive impact study that could be presented to the Regional Economic Development Council. This project provided a firsthand experience for the students to observe how technological activities can affect society and how society can affect the development of technological activities. Also, the activity represented a practical problem of meeting human needs in relation to cultural and economic consequences.

STANDARD

5 Students will develop an understanding of the effects of technology on the environment.

As with its influences on society, the impact of the use of technology on the environment can be positive or negative. Technology can clean a river or pollute it; it can clear skies or darken them. As the use of technology has grown, so too, has its potential to affect the environment — a hundred million cars have an effect that a hundred do not. It is therefore essential that all decisions about the use of technology be made with the environment in mind.

Managing resources through conservation and recycling is one of the best ways to use technology to protect the environment. The entire lifecycle of a product must be taken into account before the product is created, from the materials and processes used in its production to its eventual disposal. Optimizing a technological process can make sense both environmentally and economically because optimization minimizes waste, maximizes recycling, and conserves resources.

Increasingly, engineers have incorporated such environmental responsiveness into their designs. Many new chemical-processing technologies, for instance, have been engineered to produce less waste, as well as waste that is less toxic. Other technologies have been created to clean effluents before they are dumped into the surrounding water or air. Still others have been designed to incorporate waste materials, which would normally end up in landfills, into new products.

Yet, the careless use of technology has created negative effects through waste and by-products — the toxic sludge from a chemical plant, for example, or the emissions from automobiles. The use of technology also may cause damage simply by its presence, as when the habitat of an endangered species is displaced by a dam-created lake. There is usually little economic incentive for a company or other entity to prevent such damage from its products because the cost of the destruction is spread among the millions of people affected by it, while the cost of avoiding that damage would be borne by the company alone. Thus, a society usually creates alternate incentives via the political system and through the use of laws, regulations, and court decisions. If citizens are to participate effectively in this political process, they need to be educated about the environmental effects of technology.

GRADES

K-2

At this age, children have only a basic knowledge of the world around them. Their main focus is on their immediate surroundings and their individual lives. Thus, K-2 students should be introduced slowly to broader scale thinking that leads them to realize that their actions affect things in their individual homes, schools, and neighborhoods.

Children at this age are often encouraged to use scrap paper, used cardboard, and recycled cans when they design and make products. Because they typically demonstrate concern about the environment, younger students often cooperate enthusiastically with a school-recycling program. They should have a basic understanding that pollution can affect people and animals. They should realize that a great deal of pollution results directly from using things and then throwing them away. Teachers can boost this understanding of reuse, recycling, and pollution by teaching students the best ways to use technology.

It is important that students in grades K-2 consider whether a material, product, or system will affect the environment. To accomplish this, students could examine various materials and products to determine if they can be reused or recycled for disposal. If they conclude that an item can be reused, the students could develop ideas and plan ways to reuse the item. If they conclude that an item cannot be recycled, they could discuss an alternative plan. They could also experience hands-on recycling by designing, making, and testing a variety of containers — for cans, paper, plastic, glass, cardboard, and other materials. Afterwards, they could take their containers home and actually use them for their own recycling.

In addition to learning about the benefits of reuse and recycling, students should learn that acquiring information about materials and products is necessary in order to make decisions. This will be helpful in preparing students for lessons about the use of technology and the environment in later grades.

In order to discern the effects of technology on the environment, students in grades K-2 should learn that

A. **Some materials can be reused and/or recycled.** Materials, such as plastic or glass containers and cardboard tubes and boxes, may be reused to make useful items. Other materials, such as used newspapers, glass, and aluminum, may be recycled.

GRADES

3-5

The environment directly affects the quality of human life. Clean air and water are central to a healthy and productive life. Although the development and use of technology can solve many environmental problems, improper application may pose serious threats to the environment. It is important for students in grades 3-5 to begin to understand how technology affects the environment in both positive and negative ways. They also need to realize that it is important to look for alternative ways to protect their environment.

Students should have opportunities to explore and discuss environmental issues that are apparent locally, such as resource management and pollution. They could investigate ways that various technologies are being developed and used to reduce improper use of resources. On a broader scale, they could look at issues affecting the environment globally, such as dwindling tropical rain forests and depletion of the ozone layer. Students should use this information while learning to make decisions about the effects of technology on the environment.

The proper disposal of waste and the consistent use of recycling represent two ways to help keep the environment healthy and enjoyable for the future. It is not too soon for students to understand what happens to waste and whether it is being disposed of appropriately. Students may also explore how alternate forms of transportation can reduce pollutants that affect the environment.

In order to discern the effects of technology on the environment, students in grades 3-5 should learn that

B. **Waste must be appropriately recycled or disposed of to prevent unnecessary harm to the environment.** Biodegradable materials can be composted, and many solid materials can be recycled. It is important to reduce the amount of material going to landfills.

C. **The use of technology affects the environment in good and bad ways.** For example, the development of a mass transit system through a wooded area can improve the environment by reducing the number of automobiles traveling through it. However, such a system might cause destruction of vegetation, present danger to native animals, and compromise natural aesthetics.

GRADES
6-8

The various opportunities for understanding the lifecycle of a material or product should begin with learning about the origins of various materials and products. Students should explore how a waste product is recycled, re-used, or re-manufactured into a new product — old rubber tires being ground and re-used in road pavement or re-manufactured as soles for shoes, for example. Once they understand how products are developed, students can then examine their use and ultimate disposal. By tracing the lifecycle of a product from its inception to its disposal, students will be able to identify various points in the sequence where technology plays a role.

For example, students could study the growing of corn, tomatoes, or other garden projects. After learning details about how the soil is prepared, how the plants are fertilized, how the products are harvested, packaged, transported, and marketed, the students could examine the effects of these various processes on the environment and come up with more environmentally friendly modifications.

Unfortunately, the environment is not always friendly to humans. However, technology can be used to modify the environment. Technological products and systems have been used to improve dreadful conditions caused by natural disasters, such as earthquakes, tornadoes, and hurricanes. New construction and agricultural techniques have been developed to reduce or prevent harm to people and property.

Students should investigate how technology has affected the natural world in both positive and negative ways. Examples of positive effects of technology include wastewater treatment plants that make it possible to keep rivers, lakes, and oceans clean, and pollution-control devices that have reduced much of the destructive acid rain. On the other hand, the negative effects of technology can be observed in the depletion of the ozone layer, acid rain, deforestation, and air and water pollution. Students should learn how to objectively look at the pros and cons of a given technology in order to be informed decision makers.

Finally, students should research the potential clash between environmental and economic concerns created by technological products and systems. They could look at how the development and use of technologies sometimes cause environmental and economic concerns to be at odds. For example, students could follow the debates concerning the development of Antarctica. In the past, technological capabilities limited the use of Antarctica to a scientific reserve. Owing to recent technological advances, the future of the continent and its marine life are in question. A heightened economic interest in the potential to develop and use minerals from Antarctica is causing debate about the future use of the continent and the unknown impacts.

In order to discern the effects of technology on the environment, students in grades 6-8 should learn that

D. The management of waste produced by technological systems is an important societal issue. Recycling materials, such as glass, paper, and aluminum has decreased the waste that is sent to landfills, thereby reducing the need for new disposal sites.

E. **Technologies can be used to repair damage caused by natural disasters and to break down waste from the use of various products and systems.** New building technologies and landscaping techniques can be used to reduce the effects of earthquakes and major storms. In addition, innovative ways of reducing waste production can aid in repairing the environment. For example, the use of bacteria in sewage treatment helps to clean human waste prior to being released into rivers or lakes.

F. **Decisions to develop and use technologies often put environmental and economic concerns in direct competition with one another.** For example, decisions on the use of nuclear power, wetlands preservation, and placement of roads and highways are sometimes in direct conflict with many different viewpoints and interests.

VIGNETTE The Best Bag in Agawam

This example uses a common object, a grocery bag, for students to investigate how well it works and what the consequences of its use are for the environment. Students work in teams to gather data and to determine which grocery bag is best suited for hauling groceries safely. They also gather data regarding environmental effects and discover that the use of a technology has a direct effect on local waste management. [This example highlights some elements of Grades 6-8 *Technology Content Standards* 5, 6, 8, 9, 10, and 13.]

"Would you like paper or plastic?" This familiar question became the basis for determining the effect on the environment and a technological assessment of a common item, the grocery shopping bag. The students compared the bags by designing and using a spreadsheet to record their characteristics and their effects on the environment.

The class began the unit by brainstorming about the various characteristics of grocery bags and then selecting the most important characteristics to use in their experiment. Next, they researched different tests for evaluating each bag.

Once the tests were selected, the students divided into teams and decided which team leader positions ("Managers") would be necessary to complete the activity. The five groups included the Data Team, the Inflation Team, the Puncture Team, the Environmental Impacts Team, and the Weight Team. A sixth group, the Management Team was made up of the leaders of each of the other five teams, in addition to an overall project manager. The team leaders were assigned titles to reflect their respective teams, and the students then developed resumes and applied for the team manager positions. After reviewing the resumes and interviewing their classmates, the class selected the management team. The management team then selected the remaining students to serve as members of their respective teams and developed a timeline. The teacher served as consultant.

The Management Team selected ten stores in their town from which to gather bags. They collected the bags, along with specifications that included both the minimum and maximum dimensions.

The Data Group designed a spreadsheet to chart the test results using an ascending scale of 1 to 10. The score was assigned according to how well the bag performed on each test. The spreadsheet was designed to measure the results of the individual tests, as well as overall performance.

The Inflation Team constructed a test apparatus that allowed them to blow a blister on samples from each bag and then to compare thickness differences at set pounds per square inch (PSI). The team tested the thickness of the bags before and after inflation at different PSI and noted at what PSI the bag ruptured.

The Puncture Team designed bottles and boxes that were attached to the rims of pneumatic cylinders. After each sample bag was struck by the bottles and boxes, the Puncture Team noted at which PSI the bag ruptured.

The Weight Group suspended the bags by their handles and then filled the bags with sand in order to find out at what weight the handles would tear. After testing a couple of bags, the group decided to change their test to include the weight at which the seams separated.

The Environmental Team surveyed local consumers to determine how the bags were used after they were taken home. The students noted if the bags were recycled or trashed. The team also checked with the local waste management office for data on discarded grocery bags. All information was added to the spreadsheet for each bag.

After the tests were completed, and the scores were tallied, the bag with the highest score was declared the winner. The students then contacted the distributor of the winning bag to determine why the distributor chose to sell that particular bag. The students were surprised to learn that the bag was chosen for its looks and not its strength or recyclability. The students agreed that appearance was important because a bag with visual appeal plus durability would be reused. However, the students concluded that more study would be needed to determine if other measures could be used to limit the number of bags being wasted.

GRADES

9-12

Students in grades 9-12 need to understand the delicate balance among people, technology, and the environment. For example, conservation of natural resources, improvement of older technologies, and large-scale use of technologies are resulting in global changes. Learning to appreciate the decisions that are made to maintain a balance among society, technological use, and the environment is central to developing technological literacy.

Although many technological products or systems are developed with the good of the environment in mind, sometimes unintended side effects occur. To develop an understanding of the effect of technology on the environment, students need to study and have experiences with technological devices and systems designed to help prevent or repair damages to the environment. For example, students could design, develop, build, and control a waste-management or water-treatment system that would treat soil pollution or purify water. To reinforce this lesson, students could visit a local water treatment facility to determine how water treatment and waste-disposal practices affect the surrounding environment.

Environmental disasters are reported regularly on the evening news. Students should develop an understanding of how technological advances in landscaping and architecture are being used to mitigate such disasters. At the same time, mishandled materials and products may affect the environment and, in turn, the health and safety of people. Students could research, design, and build a model showing a cutaway view of their local terrain, complete with caverns, sand, soil, water flow patterns, ponds, and lakes. Such a model could be used to show how spilled fuels or other liquids affect watersheds and bodies of water. Students could then design and develop solutions for fixing a potential spill in their own area.

Students can determine how to evaluate their needs or wants for a product or system versus the effect that it will have on the environment. Such an evaluation involves a very complex process. For example, dams and high-powered electric lines are needed for generating power and providing service to many communities. Yet, the damming of streams and rivers and the diversion of natural water flows is damaging to many habitats and environments. Understanding the trade-offs that must be made and then making decisions accordingly helps students to recognize the positive and negative effects that can result from technological solutions.

In order to discern the effects of technology on the environment, students in grades 9-12 should learn that

G. **Humans can devise technologies to conserve water, soil, and energy through such techniques as reusing, reducing and recycling.** For example, water treatment and filtering technologies can facilitate the reuse of water; wind and water erosion can be reduced by no-till farming; and aluminum containers can be recycled.

H. **When new technologies are developed to reduce the use of resources, considerations of trade-offs are important.** Examples include the cost and limited output of photovoltaic cells to produce electricity

GRADES
9-12

mainly in remote areas and the potential long-term side effects of new drugs.

I. **With the aid of technology, various aspects of the environment can be monitored to provide information for decision-making.** The development of a wide range of instrumentation to monitor the effects of human-made gases, such as CFCs or monitor the effects of weather patterns (meteorology) and other atmospheric conditions are examples of these technologies.

J. **The alignment of technological processes with natural processes maximizes performance and reduces negative impacts on the environment.** For example, buildings can be strategically oriented to the sun to maximize solar gain, and biodegradable materials can be used as compost to make the soil more productive.

K. **Humans devise technologies to reduce the negative consequences of other technologies.** Examples include scrubbers for coal burning generation facilities, fuels that burn more clearly and, materials separation processes that aid in the recycling process.

L. **Decisions regarding the implementation of technologies involve the weighing of trade-offs between predicted positive and negative effects on the environment.** For example, the implementation of advanced transportation technologies, such as shuttles and metrorails, has had an enormous impact on the ability to travel. At the same time, roadways, urban sprawl, and automobile emissions have directly affected the environment. Indirect effects include factors such as pollution caused by manufacturing and junked cars.

STANDARD

Students will develop an understanding of the role of society in the development and use of technology.

Just as technology molds society, so too does society mold technology, shaping it in big ways and small. The most obvious influence that society wields is the vote on whether a particular product satisfies a need or want. If enough people find a product useful or desirable, it will generally be continued and developed further; if it is not accepted, it will vanish from the technological universe. In most cases this vote is tallied by the market — products that are profitable survive; others do not — but sometimes, most often for risky technologies, the decision is made politically. Nuclear power, for example, has been banned by a number of governments around the world.

Because most innovations today are developed inside organizations, organizational culture plays an important role in shaping technology — the personal computer as developed by Apple is a much different machine than the personal computer developed by IBM, for example. Government regulations, subsidies, and financial incentives can favor some technologies and be a disadvantage to others. Market forces and competition among businesses will often shape technological choice, as has happened in the battle between Microsoft and Netscape over Internet browser technology. And individuals from Henry Ford to Ralph Nader have stamped their personal marks on technology as well. Corporations will create technological demand, overshadowing needs and wants in favor of developing or increasing market value.

The values and beliefs of individuals shape their attitudes toward technology. For instance, genetic engineering is viewed by some as a way to produce more and better agricultural products at lower prices, while others see it as a possible environmental hazard and an economic threat to small farms. More generally, some people tend to be sanguine about technology, believing it to represent progress, while others view it with suspicion, arguing that its disadvantages too often outweigh its benefits.

GRADES

K-2

Young children are interested in learning about various technological products and systems and about how such things can satisfy their own needs and wants. This interest can be used to teach the lesson that, in general, such products and systems are designed with the goal of meeting individual needs, wants, or demands. For example, students might explore how the desire to see after dark and the demand for safer, more reliable sources of light led to the use of candles, oil lamps, gas lamps, and, eventually, the electric light bulb. Teachers should help students understand that the electric light bulb displaced the candle and gas lamp because it had a number of desirable characteristics: it provided a brighter, steadier, and more natural light; it was cleaner; and it was less likely to cause a fire. This example demonstrates that personal likes and dislikes and preferred characteristics help shape product development.

In order to realize the impact of society on technology, students in grades K-2 should learn that

A. **Products are made to meet individual needs and wants.** For example, people need water, so a system for providing water to the home and school was created. Because people like to play video games, computer software designers have responded with an on-going supply of new games.

VIGNETTE

The Development of the Button

This is an example of how students may approach the ideas behind how people shape the development and use of technology and relate them to the evolutionary change and history of a single product or group of similar products. [This example highlights some elements of Grades K-2 *Technology Content Standards* 4, 6, 7, and 11.]

Two people from the community, dressed in simply draped robes, visited Mr. B's elementary classroom to help introduce a unit on the development of technology. By asking carefully directed questions, Mr. B helped the children realize that the robes did not have zippers, pins, or buttons — only belts and fabric knots held the robes together. Using selected students to act as mannequins, students demonstrated how the robe was draped and secured using only the belts and fabric knots.

After the demonstration, the class reviewed and discussed information on how clothing fasteners have changed over time, including the development of buttons and buttonholes in the thirteenth century, and how the needs and wants of individuals drove those changes. To reinforce the discussion, the students experimented by making buttons out of materials representing various periods of history.

Following the making and testing of their buttons and buttonholes, the teacher and children explored why the button was so successful. Mr. B helped the children see that buttons made clothes fit better and therefore kept people warmer. The example enabled Mr. B to explain that the invention of the button helped to keep babies warm in the cold, drafty houses of the time. Thus, the button improved the survival rate of the babies — quite an accomplishment for such a simple little device.

GRADES

3-5

Students in grades 3-5 should learn that people's needs and wants have a direct influence on the development of technology. If people have no desire for a certain product or system, companies will not generally develop it. Furthermore, once people lose interest in a product or system, even one that was previously seen as a necessity, it will probably be removed from the marketplace and quickly forgotten.

In contrast, companies often encourage the demand for a product through such tactics as marketing or deliberately creating a shortage. Furthermore, because people's wants and needs are constantly changing, technology too is constantly changing. Toy sales can exemplify this principle. Many toys are made and brought to market because parents and children demand them. If the demand for certain toys drops, prices will also drop, and companies will reduce production of those items. Students should learn that when people are deciding which product to purchase, they are also influencing the rise and fall of technological development.

In order to realize the impact of society on technology, students in grades 3-5 should learn that

B. **Because people's needs and wants change, new technologies are developed, and old ones are improved to meet those changes.** Before the days of air conditioning, covered porches on homes were very popular because people could go outside to enjoy a cool breeze in a comfortable, shaded area. Now that air conditioning is widely available, new houses seldom feature open-air front porches; rather, uncovered backyard decks are favored to accommodate modern day preferences.

C. **Individual, family, community, and economic concerns may expand or limit the development of technologies.** The development of a product or system is related to the wants, interests, and acceptance of individuals. Just because a product or system could be developed does not mean it should be. Sometimes an industry is able to deliver a product or system, but because of misunderstanding or fear, a product or system is not developed. For example, the electric car, nuclear power, and genetic terminator seeds have stimulated both public mistrust and misunderstanding.

GRADES

6-8

Technology is widely recognized as causing many changes in society, but the converse is also true — society plays a critical, if not always obvious role in the development and use of technology. Students in grades 6-8 should recognize that inventions and innovations are created to help meet the demands and interests of individuals and communities. Likewise, they should have opportunities to discuss and explore various long-term research and development projects that have resulted in major impacts, such as fusion, space exploration, and genetic engineering.

Students could study how modern transportation stems from people's needs to move quickly from place to place. The invention of powerful engines and motors has helped to meet the desire for faster and more comfortable modes of travel, which in turn has affected the transportation technologies by causing an increase in the number of vehicles and roadways. The result is a spiral of ever-improving transportation technologies and an ever-increasing demand for even better ones.

In order to realize the impact of society on technology, students in grades 6-8 should learn that

D. **Throughout history, new technologies have resulted from the demands, values, and interests of individuals, businesses, industries, and societies.** The development of the typewriter helped speed the preparation of documents for many businesses, while the development of the photocopying machine revolutionized the process of duplicating documents. The typewriter and photocopying machine were followed by many other innovations including an electronic facsimile (fax) machine, and electronic mail (e-mail), which continue to change the way people correspond and keep records.

E. **The use of inventions and innovations has led to changes in society and the creation of new needs and wants.** For example, the initial creation of radios, televisions, and sound systems has led to an ever-growing demand for entertainment and information. Thus, the development of technology sometimes creates the demand.

F. **Social and cultural priorities and values are reflected in technological devices.** For example, an unenthusiastic attitude toward the use of genetically engineered foods has affected the development of this technology, yet many seed-producing companies are pressed to develop insect- and disease-resistant plants. Likewise, consumer tastes influence technological designs, such as the color and contours of household appliances. For example, new appliances are not marketed in the rounded shapes of the 1950s or the avocado green color of the 1970s.

G. **Meeting societal expectations is the driving force behind the acceptance and use of products and systems.** Whether or not a technology is accepted by society depends, first, on whether it does its job and, second, on how well it accords with various economic, political, cultural, and environmental concerns. With little regard to underlying technology, people expect buildings to provide shelter, bridges to span water, and dams to provide power and recreation.

GRADES
9-12

Technology is connected with and influenced by all of society's institutions, including economic, family, political, and educational. These societal institutions have a powerful influence on how people live, work, play, and learn. Students in grades 9-12 need to realize the influence of society on technology and how its decisions will directly affect the development of a product or system.

Students at this level begin to search for their place in a technologically complex world. They need to learn that decisions they will be asked to make can be affected by their own understanding of technology. The more opportunities they have to practice their technological thinking and decision making, the better prepared they will be to make decisions regarding outcomes of a product or system.

Students should study how public opinion and demands directly affect the marketplace. When a product or system is not regarded favorably, the developers must decide whether to continue or halt its development. Just because a product or system can be developed does not mean that it should be — acceptance or rejection by society often determines its success or failure. If companies do not consider public opinion, their products or systems can be doomed to failure, which can lead to significant financial losses. To demonstrate this concept, students could study inventions or innovations with limited success: the Edsel, named for Henry Ford's son, was considered a poorly designed automobile; after the Hindenburg disaster, production of the dirigible, a large hydrogen-filled balloon, was halted. It is extremely difficult to convince people to abandon an established product in which there has been a tremendous investment. Students should understand the importance of public opinion — those who reap the benefits have an advantage on influencing decisions regarding the development and implementation of technology.

In order to realize the impact of society on technology, students in grades 9-12 should learn that

H. **Different cultures develop their own technologies to satisfy their individual and shared needs, wants, and values.** American transportation systems are closely linked to freedom and independence, whereas other cultures might place more value on the speed and convenience associated with mass transportation systems.

I. **The decision whether to develop a technology is influenced by societal opinions and demands, in addition to corporate cultures.** The technological expertise to develop a particular product or system may be available, but if the public reaction to such development is in opposition, or if a corporation refuses to adjust to new and complex ideas, the development is most often limited or stopped.

J. **A number of different factors, such as advertising, the strength of the economy, the goals of a company, and the latest fads contribute to shaping the design of and demand for various technologies.** Sometimes these forces are consistent with one another. At other times, they may compete. The general public may or may not be aware of the influences that shape technology or of how technological development will impact the environment.

STANDARD

Students will develop an understanding of the influence of technology on history.

Technology began with very simple tools: rocks or other natural items that were modified to better serve whatever purpose their maker had in mind. As time passed, humans became more sophisticated at making tools and also learned to process raw materials into forms that did not exist in nature — bronze and iron, ceramics, glass, paper, and ink. These new materials opened the way to improve existing tools and to create whole new technologies. People learned to put individual parts together to create systems — the wheel and axle, the lever, and the bow and arrow — that could perform jobs no single item could. The division of labor allowed people to become specialists and to cooperate in making products more complicated and sophisticated than any individuals were likely to achieve on their own.

A major boost to technology came with the rise of science in the sixteenth and seventeenth centuries. Scientific knowledge opened the way to a new type of design, one not based completely on trial and error, but based partly on being able to predict how something should work even before it was built.

History has seen at least three great transformations that were driven by technology. The development of agriculture some 14,000 years ago was the first. By providing a stable food supply, agriculture allowed societies to grow and flourish, which in turn led to the first great flowering of civilization. The second transformation came in the eighteenth century with the development of the steam engine and a number of other important machines and the establishment of the first factories. These changes ushered in the Industrial Age, a time of mass production. The creation of an interconnected system of suppliers, manufacturers, distributors, financiers, and inventors revolutionized the production of material goods, making them widely available at low cost and high quality. The most recent transformation — the development of powerful computers and high-speed telecommunication networks — has taken place over the past few decades. These technologies have achieved for the field of information what the previous two revolutions did for food and material goods. The ability to store, manipulate, and transfer information quickly and inexpensively has profoundly affected almost every part of society, from education and entertainment to business and science.

Knowing the history of technology — the major eras, along with specific events and milestones — helps people understand the world around them by seeing how inventions and innovations have evolved and how they in turn produced the world as it exists today. In studying past events, one begins to see patterns that can help in anticipating the future. In this way, the study of technology equips students to make more responsible decisions about technology and its place in society.

GRADES

K-2

Studying about the history of technology in the early school grades is important because it provides students with a basic understanding of how the world around them came about. This foundation will be important as they progress through school.

Students will learn how technology has evolved from early civilizations when the first humans created primitive tools by chipping away the edges of flint stones. Making and using tools were among the first technologies; they were — and still are — a means to extend human capabilities and to help people work more comfortably. Students will realize that humans have become more than just toolmakers. Over time, people have improved their capability to create products or systems for providing shelter, food, clothing, communication, transportation, weapons, health, and culture.

In order to be aware of the history of technology, students in grades K-2 should learn that

A. **The way people live and work has changed throughout history because of technology.** Once people learned to provide shelter for themselves — first with simple huts and later with houses, castles, and skyscrapers — they were no longer forced to seek natural shelter, such as caves. The invention of the plow and other agricultural technologies, along with such simple devices as fish hooks and the bow and arrow made it easier for people to feed themselves, which freed up time for other pursuits. People's ability to communicate with one another over space and time has been improved by the use of such tools and processes as smoke signals, bells, papermaking, telephones, and the Internet.

GRADES
3-5

Throughout history, people have developed various products and systems to help in their pursuits. To understand this concept, students in grades 3-5 might study, for instance, the evolution of construction. They could trace the development of structures from the earliest people to Egyptian pyramids, Roman aqueducts, sailing ships, to modern day skyscrapers. In this way, students will come to see how the history of civilization has been closely linked to technological developments.

A variety of activities based on historical periods can help students learn how people improved their shelter, food, clothing, communication, transportation, health, and safety, and, therefore, promoted their culture. For example, to develop an understanding of the evolution of communication, students could replicate different forms of communication, starting with cave drawings and carvings and moving on to maps and charts, then to photography, and finally to graphic design. They could trace the progression of artificial light from primitive cave fires to candles, and on to gaslights and electric light bulbs, and finally neon lights, fluorescent lights, and lasers.

By the time they complete the elementary grades, students will have gained a perspective on the importance of technology through its historical development. In addition, they will have gained an understanding of the importance of tools and machines throughout history.

In order to be aware of the history of technology, students in grades 3-5 should learn that

B. **People have made tools to provide food, to make clothing, and to protect themselves.** The products and systems developed did not always work. Often many attempts and variations were tried before an idea became a reality. For example, the development of pottery stretched over 10,000 years. People learned to mix various clays to make stronger items and to fire pottery in ovens to harden the clay faster. Various containers, such as jugs, vases, and cups were designed and developed for holding things, such as water, milk, seed, and grains. Not all of the designs worked, and variations in some may be seen in every ancient civilization.

VIGNETTE

A Time Line Comparison of Communicating a Message

This example explores the history of communication and provides an opportunity for students to begin to develop and put to use their understanding of how the evolution of technologies relates to the history of humankind. [This example highlights some elements of Grades 3-5 *Technology Content Standards* 3, 4, 6, 7, 13, and 17.]

Understanding historical perspectives becomes critical when educators consider that their students have never known much technology that is more than a few years old. This point became clear to Mr. S when he realized his fourth-grade students had never heard of the railroad telegraphers' Morse Code mentioned during their lesson on westward expansion. Mr. S chose communication systems as a way to help his students explore the history of technology.

To begin the study, Mr. S provided background information, and the students conducted basic research. The students then created a time line, which depicted communication methods from prehistoric times to the present. Their time lines included such milestones as drums, messengers, whistles, mirrors, telephones, fax machines, and e-mail. The class separated into teams, and each group researched particular types of communication and then shared their findings with the other teams. The class experimented with earlier forms of communication (e.g., sending messages by foot, whistling, and using mirrors) between their school and the school down the street. After discussing the results of these basic forms of communication, the students concluded that more modern forms would be necessary for their project.

The students were given an opportunity to work with modern types of communication when their school was celebrating the networking of all 18 classrooms to the Internet. Mr. S approached his fourth graders with the challenge of testing the new system to see if it worked as well as, or better than, previous communication systems.

To do this, the class composed a message that read, "Our school has just completed networking our classrooms to the Internet, so please help us celebrate by sending back this message A.S.A.P. using the way it was sent (e.g., inter-school mail, telephone, letter, e-mail, and fax). This is a test of our new communication system. Thanks for participating in our celebration."

In teams the students sent the message to 10 other schools in the district. Each team used a different mode of communication and recorded the amount of time it took for the recipient to respond. One team used the inter-school hand-delivered mail system, another used the telephone, a third mailed letters via the postal service, a fourth e-mailed the message, and a fifth used the office fax machine.

All 10 schools replied to the message using the same form of communication from which they received it. The students then were able to compare the response rate and accuracy for each type of communication and then assess the pros and cons of each. Finally, Mr. S's students prepared reports for other classes with information about appropriate forms of communication for a given purpose (e.g., postal mail for formal invitations and e-mail for informal notes, reports). As a result of this exercise, Mr. S's class learned a great deal about the historical changes and improvements in communication, which were brought about by technology.

GRADES
6-8

In the middle-level grades, students will learn about many of the technological milestones in human history. They will recognize the ways in which technology has affected people from different historical periods — how they lived, the kind of work they did, and the decisions they made. Seeing the history of technological developments in the broader context of human history will enable students to understand how the impact of technology on humankind has changed over time.

Teachers can inspire students' curiosity about the history of technology in a variety of ways. They might, for example, have students explore various structures that provide shelter and investigate how their climate-control systems, such as heating and cooling, have made life indoors more comfortable and enjoyable. In conducting their research, students could use such sources as books, the Internet, and even older members of the community to learn about life before homes were air-conditioned and centrally heated. Once they have gathered their information, the students could present it to the class in various formats, such as building a model, making a slide presentation, or producing a video. Any number of other topics, including food, clothing, communication, transportation, weapons, and health, could also serve as the basis for such an exercise. By investigating the major inventions and innovations from various times in history, students will be able to draw conclusions about how society and culture influence technological development and vice versa.

In order to be aware of the history of technology, students in grades 6-8 should learn that

C. **Many inventions and innovations have evolved by using slow and methodical processes of tests and refinements.** For example, during the development of the incandescent light bulb, Thomas Edison and a team of 20 highly skilled technical personnel performed more than 1,000 tests before they narrowed their ideas to the one that worked. Since that first light bulb burned for 13 hours in 1879, there have been many innovations and design changes.

D. **The specialization of function has been at the heart of many technological improvements.** For example, the early steam engine was originally designed with a single chamber in which steam expanded and then was condensed — thus performing both of the two very different functions of the steam engine in the same place. Fifty years later, by isolating the functions of the cylinder and steam condenser into separate components, James Watt created a more efficient steam engine.

E. **The design and construction of structures for service or convenience have evolved from the development of techniques for measurement, controlling systems, and the understanding of spatial relationships.** For example, the purpose of Roman aqueducts was to provide a service by moving water

GRADES

6-8

from the surrounding hills to the city. The water flowed through channels, some above ground on high arches or tiers, while most were underground and were designed with a slight downward grade. Building the aqueducts required much organization, as well as an understanding of the materials and terrain.

F. In the past, an invention or innovation was not usually developed with the knowledge of science.

The introduction of science knowledge combined with technological knowledge led to a great increase in engineering and technological development. The development of a new product or system often happens in areas that have not been analyzed by science or in areas where science knowledge is being gathered alongside the technological development, such as in space programs.

4

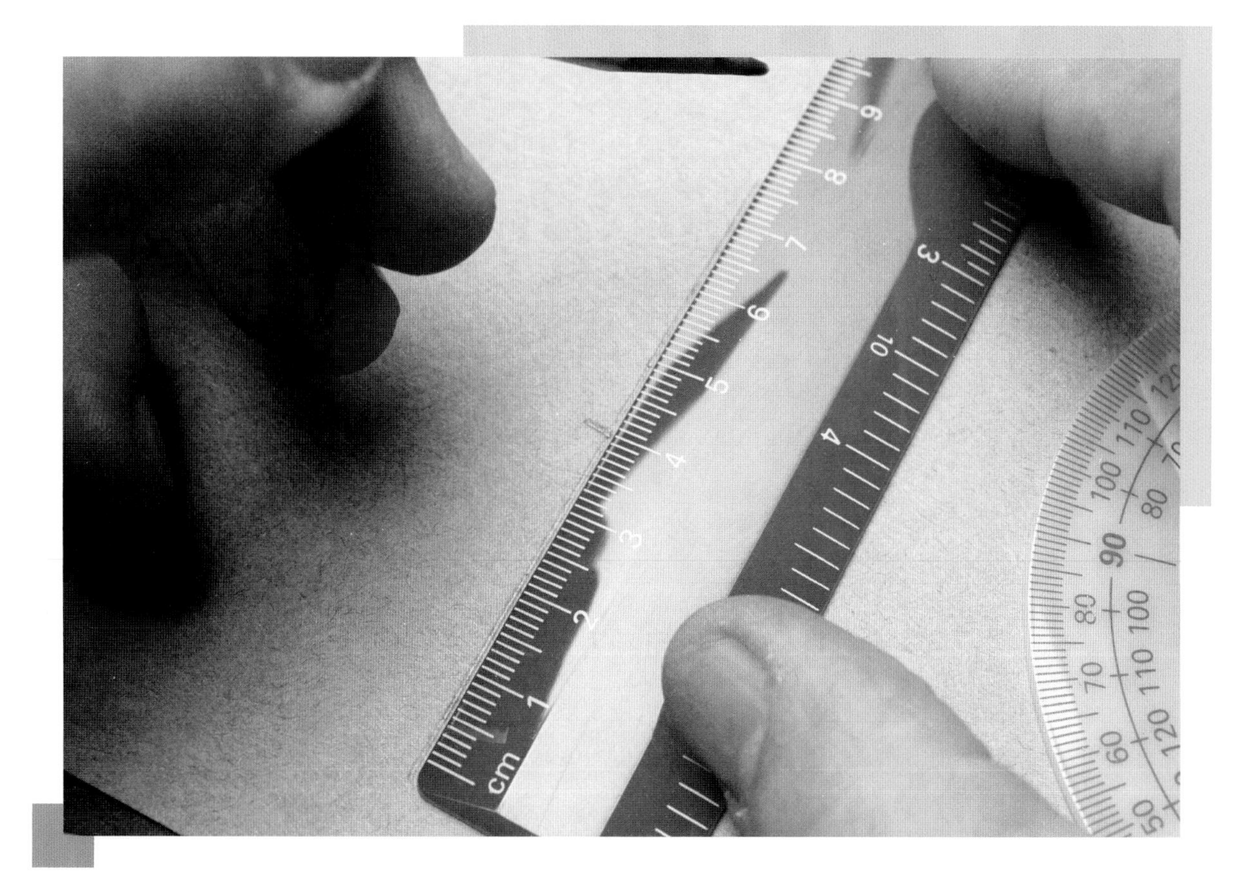

GRADES
9-12

Students in grades 9-12 should learn that sometimes technological changes are abrupt and obvious, and at other times, they are evolutionary and subtle. The effects of technological advancements can also be very powerful, irreversible, and global.

To develop an understanding of the history of technology, students at this grade level should learn about the origins and history of various inventions and innovations as they relate to particular periods of time. Historical periods have been defined and named in terms of the dominant products or systems of the time. For instance, students would learn that the Stone Age began with the development of chipped-stone tools, which later evolved into hand axes, blade tools, spears, and the bow and arrow, and that fire was also harnessed at this time. Other historical periods have been characterized by significant technological developments — the wheel, the printing press, mass production, and the computer, for example.

Without question, key developments in technology have pushed civilization forward and laid the foundation for the present high-technology era. Over the past 200 years, technological and scientific growth has become closely linked with the idea of progress. Thus, students should compare the various eras and come to understand that studying the history of technology is also studying the process of change.

Students should also understand that while history tends to be told in terms of heroes and individual inventors, in reality many people with different backgrounds have worked together and separately over time to develop technology.

In order to be aware of the history of technology, students in grades 9-12 should learn that

G. **Most technological development has been evolutionary, the result of a series of refinements to a basic invention.** For example, the development of the pencil was a long and tedious process. Engineers, designers, and technicians developed many different techniques and processes to use a variety of materials in order to develop the best pencil possible. Often a product or system will have a direct impact or dependence on another, which will affect the pace and nature of the change in one or both of them. For example, information and communication technologies have had an enormous impact on the development of the transportation system.

H. **The evolution of civilization has been directly affected by, and has in turn affected, the development and use of tools and materials.** Communication, agriculture, and transportation, for example, have evolved out of the political, economic, and social interests and values of the times. The use of electricity, farm tractors, and airplanes have enhanced safety and comfort, aided in different means of communication, and helped provide food and transportation.

I. **Throughout history, technology has been a powerful force in reshaping the social, cultural, political, and economic landscape.** The study of the history of technology helps determine possible scenarios for the future. For example, the development of the mechanical clock in the fourteenth century changed how people regarded their use of time.

J. **Early in the history of technology, the development of many tools and machines was based not on scientific knowledge but on technological know-how.** The Stone Age started with the development of stone tools used for hunting, cutting and pounding vegetables and meat and progressed to the harnessing of fire for heating, cooking, and protection. The Bronze Age began with the discovery of copper and copper-based metals. Agricultural techniques were developed to improve the cultivation of food and its supply. This period also involved the development of better ways to communicate through the development of paper, ink, and the alphabet, to navigate with boats made of timbers, and to understand human anatomy with the aid of an embalming process.

K. **The Iron Age was defined by the use of iron and steel as the primary materials for tools.** During this period, sustained technological advancement caused many people to migrate from farms to developing towns and cities. Other influential developments in this age included weaving machines and the spinning wheel, which advanced the making of cloth, and gunpowder and guns, which were an improvement over previous weapons for both hunting and protection. The wide application of new agricultural technologies, such as the sickle, the plow, the windmill, and irrigation, enabled fewer farmers to grow more food.

L. **The Middle Ages saw the development of many technological devices that produced long-lasting effects on technology and society.** This period saw the development of the waterwheel, the block printing process, paper money, the magnetic compass, and the printing press. In many ways, all of these devices are still being used today, although they have been greatly modified from their earlier designs.

M. **The Renaissance, a time of rebirth of the arts and humanities, was also an important development in the history of technology.** Leonardo Da Vinci, an Italian painter, architect, and engineer, created drawings and written descriptions of the human flying machine, a helicopter, parachutes, diving bell suit, articulated chains, a giant crossbow, and circular armored vehicles. Gunsmiths, while seeking a means to adjust their gun mechanisms, invented the first screwdriver. The camera obscura, silk knitting machines, the telescope, the submarine, the hydraulic press, and the calculating machine also were developed during this time period.

N. **The Industrial Revolution saw the development of continuous manufacturing, sophisticated transportation and communication systems, advanced construction practices, and improved education and leisure time.** Major developments of this period included the continuous-process flourmill, power loom and pattern-weaving loom, steam engine, electric motor, gasoline and diesel engines, vulcanized rubber, airplane, telegraph, telephone, radio, and television. The concepts of Eli Whitney's interchangeable parts and Henry Ford's movable conveyor added to the advances made in the production of goods. Extended free time was possible as a result of increased efficiency, and consequently, widespread education became possible because children were not needed on the farm and could stay in school longer.

O. **The Information Age places emphasis on the processing and exchange of information.** The development of binary language, transistors, microchips, and an electronic numerical integrator and calculator (ENIAC) led to an explosion of computers, calculators, and communication processes to quickly move information from place to place. Holography, cybernetics, xerographic copying, the breeder reactor, the hydrogen bomb, the lunar landing ship, communication satellites, prefabrication, biotechnology, and freeze-drying have all been major developments during this time period.

5 Design

STANDARDS

8 Students will develop an understanding of the attributes of design **p. 91**

9 Students will develop an understanding of engineering design **p. 99**

10 Students will develop an understanding of the role of troubleshooting, research and development, invention and innovation, and experimentation in problem solving **p. 106**

5 Design

Design is regarded by many as the core problem-solving process of technological development. It is as fundamental to technology as inquiry is to science and reading is to language arts. To become literate in the design process requires acquiring the cognitive and procedural knowledge needed to create a design, in addition to familiarity with the processes by which a design will be carried out to make a product or system.

More broadly, problem solving is basic to technology. Design is one type of problem solving, but not all technological problems are design problems. Technology includes many other types of problems and different approaches to solving them, including troubleshooting, research and development, invention and innovation, and experimentation.

The development of a technology begins as a desire to meet a need or want. These needs or wants could belong to a single inventor or be shared by millions of people. Once needs or wants have been identified, the designers must determine how to satisfy or solve them. The modern engineering profession has a number of well-developed methods for discovering such solutions, all of which share certain common traits. First, the designers set out to meet certain design criteria, in essence, what the design is supposed to do. Second, the designers must work under certain constraints, such as time, money, and resources. Finally, the procedures or steps of the design process are iterative and can be performed in different sequences, depending upon the details of the particular design problem. Once designers develop a solution, they test it to discover its shortcomings, and then redesign it — over and over again.

Designing in technology differs significantly from designing in art. Technological designers work within requirements to satisfy human needs and wants, while artists display their mental images and ideas with few constraints. Additionally, technological designers, such as engineers, are concerned with the usability and desirability of a product or system. As a result, efficiency is a major consideration in technological design, while the beauty or appearance of the product is often less important. In artistic design, by contrast, aesthetics and beauty are central issues, while efficiency is not. For those who appreciate them, technological designs can be viewed as works of art that showcase creativity equal to a well-crafted poem or an inspired painting. Industrial design may strike a balance between art and technology.

Over the last three decades, many countries have moved the teaching of design in technology from the periphery of the school curriculum towards its center. Because technological design involves practical, real-world problem-solving methods, it teaches valuable abilities that can be applied to everyday life and provides tools essential for living in a technological environment. Technological design also promotes teamwork as a method by which people work together to accomplish a common goal. If students know how problem-solving methods work, they can gain a better appreciation and understanding of technology. In addition, by practicing these problem-solving methods, students acquire a number of other valuable skills — performing measurements, making estimates and doing calculations — using a variety of tools, working with two- and three-dimensional models, presenting complex ideas clearly, and devising workable solutions to problems.

STANDARD

8 Students will develop an understanding of the attributes of design.

Design is the first step in the making of a product or system. Without design, the product or system cannot be made effectively. Technological design is a distinctive process with a number of defining characteristics: it is purposeful; it is based on certain requirements; it is systematic, it is iterative; it is creative; and there are many possible solutions. These fundamental attributes are central to the design and development of any product or system, from primitive flint knives to sophisticated computer chips.

Technological design is purposeful because a designer must have a goal when devising a new product or system — some function or list of functions that the product or system should perform. Without a purpose, design is no more than doodling. The design process is a system that converts inputs into outputs, or ideas into completed products and systems.

A designer or engineer is always working within requirements, such as criteria and constraints. The criteria set the parameters for a design by identifying the key elements and features of what the product or system is and what it is supposed to do. Efficiency, for example, is an important criterion in most designs. Constraints are limits on a design. Some constraints are absolute — no one can build a perpetual-motion machine, for instance. But most of the constraints that a designer works with are relative — funding, space, materials, human capabilities, time, or the environment — that must be balanced against each other and against how well the design satisfies the requirements. In order to make solutions as good as possible, the design must go through a process of optimization, with a series of adjustments being made to the design to improve its effectiveness within the given requirements. Sometimes trade-offs are made in selecting one design over another.

Technological design must be systematic. Because so many different designs and approaches exist to solving a problem, a designer is required to be systematic or else face the prospect of wandering endlessly in search of a solution. Over time, the engineering profession has developed well-tested sets of rules and design principles that provide a systematic approach to design. Design measurability, which is a key concept in the engineering profession today, is concerned with a designer's ability to quantify the design process in order to improve the efficiency. Design is not a linear, step-by-step process. Rather, it should be an iterative, or repeating process that allows designers to explore different options in a pragmatic way, become independent decision makers, and envision multiple solutions to a problem.

Technological design inevitably involves a certain amount — sometimes a great deal — of human creativity. No matter how exact the requirements or how definitive the design principles are, there are always choices to be made and there is always room for a fresh idea or a new approach. As they search for the most elegant designs that yield the best solutions, designers and engineers will depend on intuition, feelings, and impressions gained from prior experience to determine which directions to explore.

Finally, there are many possible solutions to a design problem. What may be the best solution for one situation may not be the optimum answer for another. The problem solver should look at many different solutions and determine which one (or ones) is best under the circumstances.

GRADES

K-2

For many students, the K-2 classroom will provide their first structured experience with design and technology. Starting at an early age, students should be introduced gradually to the importance of design, of visualizing objects, of translating ideas into sketches, and of using the design process to solve problems.

Research on how children learn suggests that young children's imaginations are better stimulated when they have the opportunity to work with actual materials. By working individually or brainstorming in teams, discussing their ideas, manipulating materials, and investigating how materials can be changed, students will begin to understand what design is while enhancing their imaginations.

Students at this age are creative, often demonstrating an uncanny ability to generate original solutions. In grades K-2, students need to understand that there can be several solutions to a given problem, and that some of the solutions are better for a particular situation than others. They need to be encouraged and rewarded for individual and team creativity as they formulate their own solutions.

In order to comprehend the attributes of design, students in grades K-2 should learn that

A. **Everyone can design solutions to a problem.** When searching for a purposeful solution to a design problem, many ideas should be considered, rather than looking for one right solution. For example, if asked to design a playhouse, students could brainstorm various ideas, such as using a cardboard box for the walls, building it out of plywood, or draping a sheet between chairs.

B. **Design is a creative process.** When people think about problems in order to solve them, it helps to stimulate innovation and turn ideas into action.

GRADES

3-5

In grades 3-5, students will learn that design is a useful process of planning that allows them to come up with workable solutions to everyday practical problems. Building on the foundation laid in grades K-2, students at this level should develop a more in-depth understanding of how a product or system is designed, developed, made, used, and assessed. Students should be encouraged to consider all stages when creating their designs. They should be encouraged to ask questions and provided opportunities to seek more than one solution to a given problem.

Students should recognize that positive and negative side effects are common in designing. As in life, sometimes generating a solution to one problem may create additional problems. They should have the freedom to model, test, and evaluate their designs before redesigning them. This process of continuous improvement is one of the key concepts in modern technological progress. The design process consists of a goal or purpose and is bounded by a set of requirements. Typical requirements include such things as cost, appearance, use, safety, and market appeal. In the laboratory or classroom, a specific problem, the cost of materials, and the tools that can be used are typically specified in advance. These specifications then become the exact requirements that students have to work within.

In order to realize the attributes of design, students in grades 3-5 should learn that

C. **The design process is a purposeful method of planning practical solutions to problems.** The design process helps convert ideas into products and systems. The process is intuitive and includes such things as creating ideas, putting the ideas on paper, using words and sketches, building models of the design, testing out the design, and evaluating the solution.

D. **Requirements for a design include such factors as the desired elements and features of a product or system or the limits that are placed on the design.** Technological designs typically have to meet requirements to be successful. These requirements usually relate to the purpose or function of the product or system. Other requirements, such as size and cost, describe the limits of a design.

GRADES

6-8

All humans have the ability to design and solve problems — it is a fundamental human activity. Building upon the foundation laid in grades K-5, middle-level students will strengthen their understanding of what design is and how it relates to basic human activities in everyday life.

Design is a creative process that allows people to realize their dreams and ideas for a better environment. To design is to plan, create, modify, refine, build, and enjoy. Good design turns ideas into products and systems, which, in turn, pleases and excites the users.

The design process is never considered to be final, and multiple solutions are always possible. As a result, technological problem solving differs from problem solving in other fields of study in which absolute, or "right," answers are sought.

In grades 6-8, students will learn more about the influence of requirements in the design process. At this age, students can be easily engaged in the identification of problems and opportunities they might pursue. The goal is to get them to work within reasonable requirements for a design by providing focus for their ideas. The Apollo space program, for instance, faced obvious requirements, such as cost, size, the need to withstand extreme temperatures, and a requirement for life-sustaining mechanisms for humans. These requirements forced engineers to be creative in order to put an astronaut on the moon.

Requirements encompass the factors of criteria and constraints. Learning to work with criteria and constraints is a challenge that students will face throughout life and is an important concept to understand at an early age. Some criteria questions that should be asked include: "Will it work correctly?" "Will it be effective for its design?" "Does the size appear to be appropriate?" Some constraints, which specify the limitations on the design, include: "Are the proper materials available?" "How much will this item cost?" "How much space is needed to build (or use) this product or system?" "What are the important human capabilities needed to use it?"

In order to comprehend the attributes of design, students in grades 6-8 should learn that

E. **Design is a creative planning process that leads to useful products and systems.** The design process typically occurs in teams whose members contribute different kinds of ideas and expertise. Sometimes a design is for a physical object such as a house, bridge, or appliance and sometimes it is for a non-physical thing, such as software.

F. **There is no perfect design.** All designs can be improved. The best designs optimize the desired qualities — safety, reliability, economy, and efficiency — within the given constraints. All designs build on the creative ideas of others.

G. **Requirements for a design are made up of criteria and constraints.** Criteria identify the desired elements and features of a product or system and usually relate to their purpose or function. Constraints, such as size and cost, establish the limits on a design.

VIGNETTE

Designing a Gift of Appreciation

This example uses the design process to create a gift to recognize and pay tribute to the teachers in the student's school during Teacher Apprecia-tion Week. [This example highlights some elements of Grades 6-8 *Technology Content Standards* 8, 9, 10, 11, and 19.]

After learning about the basics of materials and the design process, students were given the task of designing an appreciation gift for all the teachers in their middle school. The students outlined the design criteria and constraints, which included: the cost for each item must be less than $1.50; gifts must be designed and made in the technology laboratory; and the gift should be useful.

The class began by brainstorming possible design ideas as a group. They came up with note centers, penholders, marker racks, computer disk bins, and other gift ideas. After selecting the computer disk bins, they dis-cussed different materials to use. Next, each student worked individually researching different materials, sketching ideas, and developing a model. To estimate the cost of the various ideas, students used recent catalogs with material prices, called local dealers, and accessed information on the World Wide Web. They developed spreadsheets to calculate various combinations of costs depending on the idea.

Each student presented a model to the class. Class members evaluated the models according to the design constraints. After discussing each of the constraints, such as costs, usefulness, and aesthetic appearance, the class selected the model that would be based on their design constraints. The model was then manufactured in quantity in the technology laboratory for all teachers in the school.

GRADES

9-12

As high school students develop a greater comprehension of the design process, they will have opportunities to explore the attributes of design in greater depth. The attributes of design include the following characteristics: the design is purposeful; it is based on specific requirements; it is systematic; it is iterative; it is creative; and there are many solutions to a design problem. They will learn that seeking multiple solutions for a problem is a hallmark of the practice of technology. In order to come up with a variety of designs that offer multiple solutions to a problem, students should develop an ability to use a nonlinear approach to solving problems. The steps of the design process serve as important guideposts to help adept problem-solving students use their intuition and ingenuity to arrive at a variety of solutions. Revisiting steps in the design process allows students to view various solutions in a pragmatic way. This process provides opportunities to make adjustments to the designs. Students begin to see the systematic, yet iterative nature of the design process.

Designs typically are ill defined with no natural end to the process. In searching for the best solution, the designer redesigns, tests, refines, and remodels again and again. Sometimes an ingenious idea will allow designers to come up with a design that does exactly what they want it to do, so that they can finish the design process.

Requirements, which include criteria and constraints, are among the attributes of design to be considered. Criteria are decisions that help identify the specifications of the design. They include such factors as familial, economic, environmental, political, ethical, and societal issues that could create problems and conflicting solutions. Because requirements can compete with one another, accommodating one often results in conflicts with others. These conflicts result in trade-offs that must be considered. For example, the demand for high quality frequently competes with a desire for low cost. Because of such conflicting demands, perfect designs simply do not exist. To find the best design, students should learn to focus on as many solutions as possible.

In general, efficiency is central to the requirements for nearly every technological design. Efficiency specifies how well a given product or system performs and how close that performance is to the ideal. Optimization can help ensure that a product or system is as efficient as possible. Optimization processes include features such as experimentation, trial and error, and development.

In order to recognize the attributes of design, students in grades 9-12 should learn that

H. The design process includes defining a problem, brainstorming, researching and generating ideas, identifying criteria and specifying constraints, exploring possibilities, selecting an approach, developing a design proposal, making a model or prototype, testing and evaluating the design using specifications, refining the design, creating or making it, and communicating processes and results. The design process is a systematic, iterative approach to problem solving that promotes innovation and yields design solutions. To systematically

seek an optimum design solution, engineers and other design professionals use experience, education, established design principles, creative intuition, imagination, and culturally specific requirements.

I. **Design problems are seldom presented in a clearly defined form.** Design goals and requirements must be established and constraints must be identified and prioritized during the time when designs are being developed. Design decisions typically involve individual, familial, economic, social, ethical, and political issues. Often, these issues lead to conflicting solutions. For example, what may be politically popular may not make good economic or social sense. Based on these issues and depending on the impact of the design, certain design solutions should not be developed.

J. **The design needs to be continually checked and critiqued, and the ideas of the design must be redefined and improved.** The design process also involves considering how designs will be developed, produced, maintained, managed, used, and assessed. As a result, multiple solutions are possible. More knowledge or competing technologies cause a design to change with time.

K. **Requirements of a design, such as criteria, constraints, and efficiency, sometimes compete with each other.** When such competition happens, trade-offs occur, and the design is modified to accommodate these requirements. Different people may choose different solutions, depending on how they weigh factors.

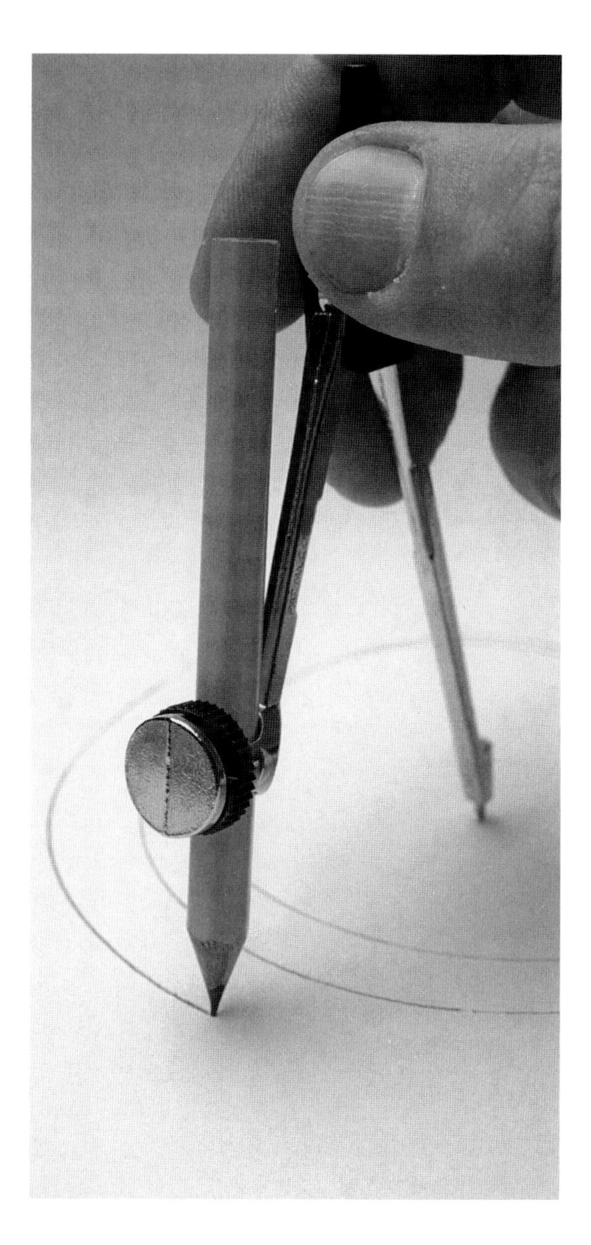

STANDARD

9 Students will develop an understanding of engineering design.

Engineers who are developing a technology use a particular approach called the engineering design process. The design process is fundamental to technology and to engineering. Also referred to as technological design, the engineering design process demands critical thinking, the application of technical knowledge, creativity, and an appreciation of the effects of a design on society and the environment.

There are many models found in literature today that attempt to describe the engineering design process. Some are linear and describe the progress as a series of steps that take place in a well-defined sequence. Many engineers, however, do not believe this model truly reflects what takes place in the engineering design process. Other models picture the engineering design process in terms of a circle with the design steps around the circumference or as a spiral. These models try to represent the iterative nature of the engineering design process as well as to indicate that the steps of the process do not have to begin in any prescribed sequence. Although the engineering profession has not come to a consensus on which model best describes the process, they do agree on several steps that should be included when describing it. These steps do not have to be performed in a set order, but rather they should be used by designers in ways that their intuition tells them is best suited for solving the problem at hand. The environment in which the engineers design should be open and encourage creativity.

One step in the engineering design process is identifying the problem. Another step is generating ideas by using such techniques as brainstorming and conducting research. The requirements of the problem should be identified, and the designer should explore possibilities for solving the problem and then select approaches that may lead to solutions. To help evaluate the solutions, models and prototypes can be built and tested, and the results can then be used to determine how well the solutions meet the previously identified requirements. The solution must constantly be refined as information is gathered through feedback and new ideas are generated. It may be necessary to retrace a number of steps in order to iteratively refine the design solution before the optimum one is selected. One of the final steps in the engineering design process is to build or construct the actual product or system in order to determine if it works. Once the designer is pleased with the solution, the final product or idea can be produced and marketed.

GRADES
K-2

Children at this age enjoy doodling, sketching, and building simple things. In grades K-2, they will learn about and be able to apply these abilities, and others, as they are introduced to the engineering design process. Students will understand that the engineering design process is a method used to solve problems. All the products and systems they see around them must have been designed and made — from the fork they eat lunch with, to the toys they play with, to the clothes they wear.

The engineering design process helps give structure to creative and innovative thinking. The process includes a number of steps, which are appropriate for young children to learn. In a very simple form, the steps of the engineering design process include identifying the problem, looking for ideas, developing solutions, and sharing solutions with others. Because students at this age are focused on their immediate environments, they should be given problems that relate to their individual lives, including their interactions with family and school environments. Looking for ideas, or researching, can take many forms, including reading books and talking to others. Another method for generating new ideas is for students to investigate things they currently use and search for ways to improve them. As a result of their research, students often will develop several solutions.

As they use the engineering design process, students should communicate their ideas and solutions to classmates, teachers, and family and community members using sketches, models, and verbal descriptions. Through this communication process, they will be able to reflect on their progress, as well as to receive ideas from others.

In order to comprehend engineering design, students in grades K-2 should learn that

A. **The engineering design process includes identifying a problem, looking for ideas, developing solutions, and sharing solutions with others.** In the design process, there are many solutions to a problem, with some being better than others. Each design can be made better by refining it.

B. **Expressing ideas to others verbally and through sketches and models is an important part of the design process.** A sketch, which typically describes the appearance of a product, can help put ideas into a form that can be used to communicate with others. Sketches are more efficient than words for conveying the size, shape, and function of an object, while models are effective in imparting a three-dimensional realism to a design idea.

VIGNETTE

Can You Help Mike Mulligan?

This vignette is an example of using childhood literature in the study of technology. It uses a problem presented in a book to create a classroom learning activity. This example can be used in the K-2 classroom to stimulate interest and motivate students toward technology and literature. [This example highlights some elements of Grades K-2 *Technology Content Standards* 3, 9, and 10.]

Virginia Lee Burton's book, *Mike Mulligan and His Steam Shovel*, provided a problem-solving challenge for Mr. C's second grade class. After reading this story to the point that Mike Mulligan realizes he didn't leave a way out of the cellar hole, Mr. C asked his students to identify the problem. The students recognized that Mike Mulligan and the steam shovel were stuck in the hole. Mr. C then engaged the class in a brainstorming session to generate various methods that Mike Mulligan could use to get the steam shovel out of the hole.

After compiling a list of their ideas on the board, Mr. C divided the class into teams of three to four students. Each group was given a tub of wet sand with a hole dug in it and a miniature steam shovel in the bottom of the hole. They were also given a box of materials that included such items as spools, straws, string, wire, Popsicle sticks, yarn, paper clips, clay, glue, tape, and rubber bands. He challenged them to use the problem-solving skills that they had been learning in class to find a method to get the steam shovel out of the hole without touching it.

The students then went to work as a group designing various methods that would allow them to solve the problem. The first solutions didn't work for most of the groups. After evaluating why their ideas didn't work, some groups decided to redesign their solutions while others came up with new solutions. Mr. C offered the students resource books that contained suggestions for various solutions.

Once the students had generated a solution and built a model of it, they sketched a picture and labeled the simple machines that were involved in the method. Each group then wrote a new ending to the story based on the machine that their group developed.

The entire class gathered for the presentations. The groups demonstrated the method and the machine they had created and read their stories to the class. After the presentations were finished, Mr. C read the end of the story to the class.

GRADES

3-5

Students in grades 3-5 should build upon what they have learned in K-2 about the engineering design process by adding several additional steps. They should realize, for example, that the purpose of the engineering design process is to convert ideas into finished products and systems. Sometimes a design results in physical products (e.g., a sewing machine, a bridge, or a car), and at other times, a design may result in processes (e.g., how to use a computer program, how to make a drawing, or how to bake brownies).

The engineering design process as understood by students in grades 3-5 includes defining the problem, generating ideas, selecting a solution, making the item, evaluating the outcome, and presenting the results. As the students work with the engineering design process, it is important that they realize that these steps do not have to be completed in a set sequence. Rather, they should be completed in any sequence that will produce the best results.

Each step of the engineering design process involves students obtaining certain information and developing specific skills. When generating ideas, for instance, students should be encouraged to be creative and to consider thoughtfully all ideas. Once they select the solutions, students should make sketches and drawings of what they will look like. Then, using available resources, they should create or make their solutions and evaluate them. Evaluation is a back-and-forth process of assessing the performance of solutions and then using that information to fine-tune and improve them. Once the students have finalized their solutions, they should present what they have learned to others in their class, to the teacher, and to other members of the school and community. In this communication process, students should describe not only what went well, but also some of the obstacles that they encountered in the engineering design process.

In order to comprehend engineering design, students in grades 3-5 should learn that

C. **The engineering design process involves defining a problem, generating ideas, selecting a solution, testing the solution(s), making the item, evaluating it, and presenting the results.** At the beginning of this process, it is important that students gather as much information about the problem as they can find. This "wide open" consideration of all ideas will help as they seek the best solution for their problem.

D. **When designing an object, it is important to be creative and consider all ideas.** The design process can unlock creative thinking and turn ideas into reality. Having a lot of ideas gives the designer many possibilities to draw from and use.

E. **Models are used to communicate and test design ideas and processes.** Models are replicas of an object in three-dimensional form. Models can be used to test ideas, make changes to designs, and to learn more about what would happen to a similar, real object.

GRADES

6-8

When learning the engineering design process the teacher and students first need to define the problem. Once the problem is determined, brainstorming becomes an important group problem-solving technique for generating as many ideas as possible. It allows for creative input from a number of people and encourages everyone to speak without fear of their ideas being judged or belittled. The more ideas an individual can draw from, the better the chances that an optimum solution can be found. After the initial brainstorming session is completed, the group should determine which of the ideas suggested are the most appropriate. These ideas should then be researched in more depth. The designer also needs to specify constraints and identify criteria in order to establish the requirements of the design. Throughout the iterative process, alternative solutions should be considered.

At this point, an approach for solving a problem should be selected, and a design proposal should be developed. A design proposal is a written plan that specifies what the design will look like and what resources are needed to develop it. It can be communicated through various forms, such as sketches, drawings, models, and written instructions. Models allow a designer to make a smaller version without having to invest the time and expense of making the larger item. Physical, mathematical, and graphic models can also be used to communicate an idea.

After an idea has been developed, it is important to test and evaluate the design based on the requirements. This testing and evaluating process leads to the refinement and improvement of the design. Next, the refined design is developed and produced. This may involve making one or more items.

In order to comprehend engineering design, students in grades 6-8 should learn that

F. **Design involves a set of steps, which can be performed in different sequences and repeated as needed.** Each design problem is unique and may require different procedures or demand that the steps be performed in a different sequence. In addition, engineers and designers also have their preferences and problem-solving styles and may choose to approach the design process in different ways.

G. **Brainstorming is a group problem-solving design process in which each person in the group presents his or her ideas in an open forum.** In this process, no person is allowed to criticize anyone else's ideas regardless of how inane they may seem. After all of the ideas are recorded, the group selects the best ones, and then further develops them.

H. **Modeling, testing, evaluating, and modifying are used to transform ideas into practical solutions.** Historically, this process has centered on creating and testing physical models. Models are especially important for the design of large items, such as cars, spacecraft, and airplanes because it is cheaper to analyze a model before the final products and systems are actually made. Evaluation is used to determine how well the designs meet the established criteria and to provide direction for refinement. Evaluation procedures range from visually inspecting to actually operating and testing products and systems.

GRADES

9-12

An engineer is, in essence, a problem solver who uses the engineering design process to solve problems. The task of an engineer is more than simply designing a product that works. He or she must consider many other factors, such as safety, environmental concerns, ethical considerations, and risks and benefits. In the design process, it is vital that people with different interests and expertise work together when devising solutions to a problem. These diverse individuals bring various perspectives to the solution of a problem.

In educating students about engineering design, a teacher must stimulate the curiosity of the students so that they become interested in the design process and motivated to learn more about it. The students should have many opportunities to design so that they will develop an in-depth understanding of this important process.

Students in grades 9-12 are introduced to two more concepts in the engineering design process: making prototypes and using design principles. A prototype is a working model that is conceived early in the design process. Prototypes provide a means for testing and evaluating the design by making observations and necessary adjustments. Computer prototypes allow design solutions to be tested in virtual settings. Students are also introduced to design principles, such as balance, proportion, function, and flexibility. These principles are universal across all types of design and establish the rules that designers use to create designs that are pleasing to the eye and functional to use. Design principles also are used to evaluate existing designs and to collect data.

There are many factors that influence a design, including how safe the design solution will be. Reliability is another concern in the design process, as is quality control. Environmental concerns, as well as how well the solution can be produced (manufactured), must be taken into account when designing solutions to technological problems. After designed products or systems have been created, it is important to maintain and repair them. This becomes a consideration which must be incorporated into the design. Finally, human factors engineering, sometimes referred to as ergonomics, is another significant concept that is applied to many designs. Human factors engineering is concerned with how a design can be used to modify tools, machines, and the environment to better fit human needs. For example, ergonomically designed chairs are easier to sit in and provide positive support to the human body.

In order to comprehend engineering design, students in grades 9-12 should learn that

- **I. Established design principles are used to evaluate existing designs, to collect data, and to guide the design process.** The design principles include flexibility, balance, function, and proportion. These principles can be applied in many types of design and are common to all technologies.

- **J. Engineering design is influenced by personal characteristics, such as creativity, resourcefulness, and the ability to visualize and think abstractly.** Individuals and groups of people who possess combinations of these characteristics tend to be good at generating numerous alternative solutions to problems. The design process often involves a group effort among individuals with varied experiences, backgrounds, and interests. Such collaboration tends to enhance creativity, expand the range of possibilities, and increase the level of expertise directed toward design problems.

K. **A prototype is a working model used to test a design concept by making actual observations and necessary adjustments.** Prototyping helps to determine the effectiveness of a design by allowing a design to be tested before it is built. Prototypes are vital to the testing and refinement of a product or system with complicated operations (e.g., automobiles, household appliances, and computer programs).

L. **The process of engineering design takes into account a number of factors.** These factors include safety, reliability, economic considerations, quality control, environmental concerns, manufacturability, maintenance and repair, and human factors engineering (ergonomics).

STANDARD

10

Students will develop an understanding of the role of troubleshooting, research and development, invention and innovation, and experimentation in problem solving.

Engineering design is a major type of problem-solving process, but it is not the only one. There are many other approaches that are used in solving either formal (well-defined) or informal (ill-defined) problems. Trouble-shooting is a specific form of problem solving aimed at identifying the cause of a malfunctioning system. Often the problem can be traced to a single fault, like a broken wire, a burned-out fuse, or a bad switch. Good troubleshooters are systematic in eliminating various possible explanations as they focus on the source of the problem.

As a problem-solving method, research and development (R&D) is much broader in scope than troubleshooting. After something has been conceived, it can take considerable time for teams of people to refine and work the bugs out before it becomes a product ready for market. If there are flaws in the design, these need to be researched, analyzed, redesigned, and corrected. Unlike troubleshooting, R&D tends to address a wide range of issues concurrently. The product must work. It must be reliable, safe, and have market appeal. Sometimes, questions about its value to society or potential harm to the environment must be researched and addressed.

Invention and innovation are among the most open-ended and creative problem-solving approaches. Unlike other forms of problem solving that deal with things already in existence, invention launches into the unknown and the untried. Invention is the process of coming up with new ideas, while design is concerned with applying these ideas. On the other hand, an innovation is an improvement of an existing product, system, or method of doing something. Creativity, in addition to an ability to think outside the box and imagine new possibilities, is central to the processes of invention and innovation. All technological products and systems first existed in the human imagination.

Experimentation is the form of technological problem solving that resembles most closely the methods that scientists use. Using methods that are similar to the scientific approach, technological problem solvers apply iterative processes to experiment on technological products and systems. For example, performing hardness tests on various metals may be needed before using those metals to make tools. Another example is testing airplanes in various situations to see why similar models crashed. Because the goals of technologists and scientists differ, their approaches to work also differ. Scientists use experiments to gain a better understanding of the natural world. Technologists, on the other hand, use experiments to understand and change the human-made world. Quality control should be used in the process of experimentation to assure that a desired standard is met.

These different types of problem solving are not always easy to distinguish from one another. Sometimes they go on at the same time as teams focus on very large problems. In addition, some problems require the expertise of both science and technology in order to find solutions.

GRADES

K-2

In the early grades, students will learn some of the basic approaches to problem solving. The design process, one approach to problem solving, was discussed in the previous two standards. Other approaches to problem solving can also be introduced at this level. For example, when a product or system quits working, troubleshooting can be used to isolate and correct the problem. Students should be introduced to troubleshooting by learning how to correct problems with simple systems. For example, they could determine and correct a problem with a flashlight that does not produce light. Using a systematic process, students can determine whether the bulb, batteries, or the switch was the source of the problem.

Young students also can be inventive. Students at this level enjoy the challenge of inventing something new for a given purpose. Students should be taught the best ways to ask questions in order to get accurate and timely information. Additionally, they should gain the ability to observe technological processes, products, and systems to gain a firsthand knowledge on how things function. Teachers should create a non-threatening working environment that encourages students to come up with ideas.

Another important concept for children to learn is that malfunction and failure are common in technological products and systems. With proper maintenance, many of these products and systems can be made to last longer. When they do fail, they often can be repaired. At other times, however, the products and systems cannot be fixed and must be discarded.

In order to be able to comprehend other problem-solving approaches, students in grades K-2 should learn that

A **Asking questions and making observations helps a person to figure out how things work.** One of the best ways to learn is through asking simple questions: "How do these two parts fit together?" or "What tool do we need to fix the bicycle?" Another important way of learning is to look at something and try to figure out how it works.

B. **All products and systems are subject to failure. Many products and systems, however, can be fixed.** Some stop working because they are old, and others because a part wears out. Troubleshooting helps people find what is wrong with the product or system so that it can then be fixed. Products and systems need to be maintained in order to keep them in good operating order.

GRADES

3-5

In grades 3-5, students should build upon their problem-solving abilities that were developed in earlier grades. They should be challenged to troubleshoot more complex systems that do not work.

Invention and innovation can be especially exciting for students in grades 3-5. For example, students could be challenged to invent a toy for preschool children. To learn about innovation, students could be challenged to modify an existing toy in order to improve upon its design or purpose.

Experimentation is also an important part of technology. During the fourth and fifth grades, students will be introduced to it in their science lessons. Experimentation in technology can be demonstrated through the search for solutions to technological problems. For example, the problem is identified, a hunch (hypothesis) about the source of the problem is generated, tests are conducted, and data is gathered. These data often reveal the nature of a problem, which helps in knowing the proper course of action to take in order to solve it.

In order to be able to comprehend other problem-solving approaches, students in grades 3-5 should learn that

C. **Troubleshooting is a way of finding out why something does not work so that it can be fixed.** Troubleshooting involves a logical and orderly process of discovering what the problem is in a part or system.

D. **Invention and innovation are creative ways to turn ideas into real things.** Technology starts with invention and is improved through innovation. Inventions are new things, while innovations change things that already exist.

E. **The process of experimentation, which is common in science, can also be used to solve technological problems.** Typically, experimentation includes testing something under controlled conditions in order to improve or change it.

VIGNETTE

Navigational Technology

This vignette presents some problems to solve related to transportation technology. Specifically, students are encouraged to invent new ways and innovate older methods to navigate while on a ship. [This example highlights some elements of Grades 3-5 *Technology Content Standards* 3, 10, and 18.]

Historical examples of how individuals used problem-solving skills to solve technological problems can provide opportunities for students to learn a variety of principles from several fields of study. During a unit on exploration, Ms. H assigned her class to read *Pedro's Journal*, a novel by Pam Conrad. This book provided the backdrop to investigate different examples of navigational technology.

Students discovered that with relative ease, sailors could determine their latitude by measuring the angle of elevation of the North Star (Polaris). Ms. H then introduced the concepts of angles and a global-grid coordinate system. In the classroom, students used protractors and measuring angles to construct working astrolabes. Using multimedia software, the apparent motion of the stars was simulated. It became clear to the students that Polaris was the only star in the Northern Hemisphere that could be relied upon to remain in a constant position relative to the observer.

An accurate star chart of the circumpolar constellations was posted on the ceiling of the classroom. Students then used problem solving to navigate their way around the room, determining the "latitude" of their desk, or whether the pencil sharpener was at more northern or more southern latitude than the teacher's desk. In a discussion, the class came to the conclusion that the astrolabes were only able to specify the latitude and that several points around the room seemed to share the same latitude. It was clear that the early mariners needed to develop additional technology to position themselves precisely while at sea.

The students then were assigned teams to develop navigational equipment that could have solved this problem and allowed the mariners to precisely position themselves on the open seas. The groups brainstormed different ideas and conducted research in the library and on the Internet before they selected the best idea to pursue. With the help of their teacher, each group made a model of its equipment and tested it to determine how well it worked. The groups later presented their findings to their classmates and demonstrated their equipment.

In the novel, the students read about sailors measuring speed by putting a rope over the stern and counting the knots as the rope ran over the gunwale. The class talked about how inaccuracies in this method gave Columbus the opportunity to easily deceive his crew into believing that they were closer to home and further east than they really were.

Through studying the development of navigational technology, the class learned about various methods of problem solving, measurement, angles, grid coordinate systems, and precision versus accuracy; astronomy; speed, time, and distance calculations; researching, designing, developing, and testing; Western history; and reading comprehension.

GRADES

6-8

At the middle level, students will work with solutions to more complicated and demanding technological problems. By the time students enter the middle grades, they should be able to distinguish different kinds of problems. For example, they should realize that designing something using a set of requirements requires a different problem-solving process than determining why a device does not work. In addition to design, students at this level should expand their knowledge about problem solving to include troubleshooting, invention and innovation, and experimentation. Different kinds of problem solving require different abilities, knowledge, attitudes, and personalities. Sometimes success with problem solving comes down to self confidence and trusting one's ability and instincts. Because people are unique, with different strengths to offer, they differ in their ability to solve various types of problems. As a result, teamwork becomes important. Teamwork allows individuals to pool their strengths in order to arrive at better solutions to problems.

Inventors tend to be creative and have excellent imaginations. They often demonstrate the ability to see possibilities that others miss. By contrast, trouble-shooting almost always requires specific knowledge. To figure out why an automobile does not start requires specific knowledge about automobile systems. Without the right kind of knowledge, many people resort to inefficient and ineffective practices and oftentimes still fail to find the cause of the problem. Experimentation is one of the most formal types of problem solving, requiring a person to follow an established set of procedures.

In order to be able to comprehend other problem-solving approaches, students in grades 6-8 should learn that

F. **Troubleshooting is a problem-solving method used to identify the cause of a malfunction in a technological system.** These kinds of problems typically require some type of specialized knowledge. For example, knowledge about how a derailleur works is needed in order to find out why a bicycle does not shift properly. Once the cause of the problem has been identified, the next step is to repair and test it.

G. **Invention is a process of turning ideas and imagination into devices and systems. Innovation is the process of modifying an existing product or system to improve it.** All technological refinement occurs through the process of innovation.

H. **Some technological problems are best solved through experimentation.** These include experimentation with technological products and systems. This process closely resembles the scientific method. The difference between these methods is the goals that each pursue. The goal of science is to understand how nature works, while the goal of technology is to create the human-made world. In both cases, the process is systematic and involves tinkering, hypothesizing, observing, tweaking, testing, and documenting.

GRADES
9-12

In addition to learning about troubleshooting, invention and innovation, and experimentation, students at this level will learn to engage in research and development (R&D). R&D is a goal-oriented process in which designs, inventions, and innovations are researched and refined to address a range of objectives and concerns. These concerns can be functional (e.g., making it work better), economic (e.g., giving it market appeal), and ethical (e.g., making it safer). R&D pursues answers to unknown questions that need to be solved before a design can work. Products and systems that are being prepared for the marketplace should almost always go through an extensive period of R&D involving teams of people with wide-ranging expertise.

During R&D, several different problem-solving strategies are often applied to the same problem, either concurrently or in sequence. In the prototyping stage of the design process, for instance, it is often necessary to use troubleshooting to get the prototype to work.

Students should also realize that knowledge from many fields of study is required to solve technological problems. Just knowing about technology in order to solve a technological problem is not sufficient. For example, it is impossible to know exactly what knowledge will be needed to build and put people on the international space station. Psychologists and physiologists, for instance, will help design the proper ergonomics of the space station. Dieticians will specify diets; medical doctors will focus on health issues; and economists will concentrate on costs. Frequently, solutions to difficult problems are found when someone with a very different kind of knowledge or perspective injects that thinking into a given situation.

In order to be able to comprehend other problem-solving approaches, students in grades 9-12 should learn that

I. **Research and development is a specific problem-solving approach that is used intensively in business and industry to prepare devices and systems for the marketplace.** Research on specific topics of interest to the government or business and industry can provide more information on a subject, and, in many cases, it can provide the knowledge to create an invention or innovation. Development helps to prepare a product or system for final production. Product development of this type frequently requires sustained effort from teams of people having diverse backgrounds.

J. **Technological problems must be researched before they can be solved.** When a problem appears, it is first necessary to learn enough about it to decide the best type of problem-solving method.

K. **Not all problems are technological, and not every problem can be solved using technology.** Technology cannot be used to provide successful solutions to all problems or to fulfill every human need or want. Instead, some problems do best with non-technological solutions. For example, recycling to reduce pollution and conserve resources is a behavioral solution to a technological problem. In the area of healthcare, healthy living practices, such as good

GRADES
9-12

nutrition and regular exercise, can often prevent and solve problems that surgery and medications cannot.

L. **Many technological problems require a multidisciplinary approach.** Depending on the nature of a problem, a wide range of knowledge may be required. For example, the research and development of a new video game could benefit from knowledge of physiology (e.g., reaction times and hand-eye coordination) as well as psychology (e.g., attention span and memory).

5

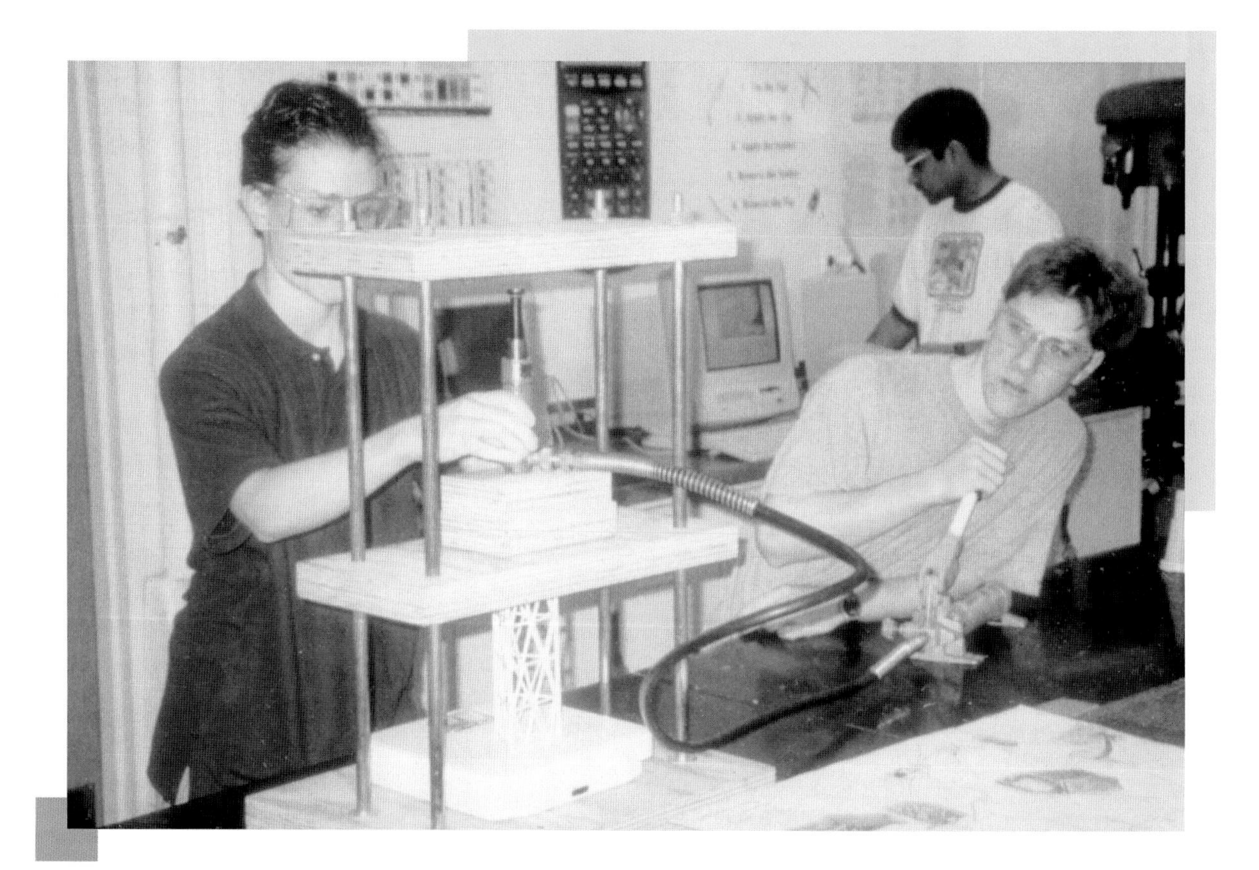

6 Abilities for a Technological World

STANDARDS

6 Abilities for a Technological World

As technologies become more powerful and more useful, they generally become more complex.

The quill pen becomes the typewriter that becomes the word processor. The horse and buggy becomes the Model T that becomes the family minivan. As a technology grows, so too does the commitment required to understand and be comfortable with the technology. Every farmer's child could fix the horse and buggy when required, and most Model T owners soon learned how to keep it running. But when the family minivan breaks down, all but a few will haul it into the shop and let a specialist take care of it.

With each passing decade, technological knowledge becomes more specialized and widespread. To a certain extent, this pattern is inevitable. However, the average citizen should not automatically delegate the ability to use and work with technology to members of the high-tech community who develop and operate it. On the contrary, because the use of technology represents such an important part of life, everyone needs a broad understanding of what it is, how it is developed, how it works, and how to make intelligent decisions about it.

Technological literacy can be defined as having the ability to use, manage, assess, and understand technological products and systems. This ability, in turn, demands certain mental tools, such as problem solving, visual imaging, critical thinking, and reasoning. The development of these capabilities is central to technological literacy. These various skills can be developed in students through such activities as modeling, testing, troubleshooting, observing, analyzing, and investigating.

The content standards in this chapter involve the development of important abilities for a technological world, which include applying the design process, using and maintaining technological products and systems, assessing products and systems, and others.

STANDARD

11 Students will develop abilities to apply the design process.

Very few products and systems today are developed by trial and error or come about by accident. Instead, almost any technology that a student encounters is the result of a systematic problem-solving design process that transformed an idea into a final product or system. This design process involves an in-depth understanding of the problem and resources available, an exhaustive search for solutions, and an extensive evaluation and refinement procedure. The design process is the foundation for all technological activity.

Most people think that the design process should be left to engineers or designers, but, in fact, everyone has the ability to design. By following the iterative steps of the design process, anyone, from first graders to the elderly, can learn to design.

The design process requires the use of a variety of strategies, such as problem solving, creative thinking, visual imagery, critical thinking, and reasoning. It also requires hands-on abilities, such as measuring, drawing, sketching, working with computers, and using tools. Quality control is also an essential factor in the design process to maintain a desired standard of quality in what is being designed, as well as arriving at the optimum solution.

Standard 8 asks that students know the attributes of design, and Standard 9 calls for them to know the steps in a design process. Besides knowing the attributes of design and being able to articulate the steps of the design process, technologically literate students must also be able to apply the process. Thus, where the previous standards dealt with what students should know or understand about the design process, Standard 11 deals with the application of the design process. These knowledge and process standards are connected — the design process cannot truly be understood without the opportunity to apply it. Conversely, students cannot successfully apply the design process without a cognitive understanding of what they are doing.

The skills required by the design process are valuable in and of themselves. In the process of developing those abilities, students will also attain firsthand experience about the transformation of ideas into solutions, which in turn will make them more comfortable with technology.

GRADES

K-2

In grades K-2, students are active, energetic learners who enjoy investigating and exploring the world around them. In addition to encouraging their students to be creative and imaginative, teachers should establish an environment in which students can design, develop, test, and communicate their ideas. The use of the design process provides an opportunity for students to generate and express ideas in a supportive and structured setting.

The problems that students identify for the design process should come from everyday experiences. They should be encouraged to explore the world around them as it relates to their personal needs and wants, and teachers are urged to select appropriate problems for children at this level. Once the teacher and students have identified a problem, the students should begin developing potential solutions.

After they have selected a solution, students should build or construct it to demonstrate the design idea. It is important for the teacher to instill in the students the value of safety when using tools and materials. The building process will give students valuable experience with various skills for handling materials, such as measuring, marking, cutting, shaping, assembling, and combining.

While generating solutions, students should be given many opportunities to see how things that are already made can be improved. The students should work with various types of materials and come up with ways to test different materials in order to determine their properties. For example, they could test newspaper and plastic to determine which is more waterproof or heavier. Another experiment might test which materials could easily be shaped (cut with scissors) or fastened (glued). Based on the information gathered, the students then could choose which materials they would like to use. In addition, to help them in their search for solutions, students could brainstorm with their classmates and receive input from their teacher. Throughout the entire design process, students should communicate their ideas, solutions, and results to others both in and out of the classroom.

As part of learning how to apply design processes, students in grades K-2 should be able to

A. **Brainstorm people's needs and wants and pick some problems that can be solved through the design process.** These problems can come from everyday problems — family or school dilemmas, for example.

B. **Build or construct an object using the design process.** Using the design process, students can build or construct it in three-dimensional form. This could include building a scaled-down model of the object.

C. **Investigate how things are made and how they can be improved.** A sense of innovation can be developed through this investigation.

VIGNETTE

Building Something to Float

This vignette presents a process used in an elementary school class to design a boat using certain requirements. It provided a simulated experience to a real-world building process. [This vignette highlights some elements of the Grades K-2 *Technology Content Standards* which provide connections with Standards 8, 9, 10, 11, 19, and 20.]

Ms. B gave her second-grade students equal size pieces of wrapping paper, aluminum foil, and wax paper. They also received clay beads, golf balls, small pebbles, marbles, and a tub of water. Ms. B then asked her students to design and make one boat that would hold nine objects and float on the water for 10 minutes.

Before making the product, the students, with guidance from their teacher, thought about the different design options, such as which material should be used and what would be the most appropriate shape. The students then selected a shape, chose their materials, and made their boats. They tested their boats to determine which shape would hold the nine objects. After that experiment, some students tried other materials to see if they worked better, while others added more objects to determine at what point their boats would sink.

After the test, the students identified which products held the loads. Discussion questions included: Why did one shape float and another sink? Which was the optimum design and why? What other materials would have worked? What kinds of similar products have people made at different times and places? Based on the discussions and responses to these questions, students were asked to reflect on them and draw some conclusions.

GRADES

3-5

In grades 3-5, students will expand their understanding of and ability to use the design process. As stated before, the design process begins with identifying a problem that can be solved through the use of technology. The problem should be one that interests the students personally, although it might not involve their direct needs or wants.

Students will next generate ideas for solving the problem. They should be encouraged to recognize that some ideas for solving problems will differ from their own. At this stage, students should collect information to help in identifying requirements for the design problem and for developing the solution. The more information they can collect, the more likely they are to find a working solution. This collection process starts with formulating questions to guide the search. Techniques for finding answers to these questions include searching the Internet, interviewing experts, reading books, and looking at similar products or systems. Students also will experiment with various materials, tools, and resources in order to select the best one for their needs.

After this information has been gathered, students will select the best possible solutions and then create a design using sketches and drawings. Students also must learn to use tools and machines safely and effectively. If they are making a physical product, for example, they may need to perform the basic processes of separating, forming, and combining materials in order to complete their task. An example would be designing and constructing a paper house. In this case, they could use computers, markers, colored paper, scissors, and paste to design and build their house so that it would be attractive and functional.

Next, students would test and evaluate the effectiveness of their solution. During this phase, they should be encouraged to reflect on the requirements that have been identified. It is also important that they address central issues: Does it work well? Does it meet the criteria that were established earlier in the design process? How effective is the design in solving the problem? After answering these questions, the students should improve their solution by redesigning it.

It is important that students learn that applying the design process involves iteration. They should learn how to use repetition and recurrence — "do it over again" — techniques to obtain the desired solution to a problem. Throughout the entire design process, students should work to improve the designed solutions. Additionally, they should communicate with other members of the class by sharing their ideas and accepting input. When the students believe they have a good solution, they should give a presentation to the class, the teacher, parents, and possibly the community. Communicating what they have done will reinforce and strengthen what they have learned.

As part of learning how to apply design processes, students in grades 3-5 should be able to

- **D. Identify and collect information about everyday problems that can be solved by technology, and generate ideas and requirements for solving a problem.** In collecting information, it may be necessary to use printed material (books or magazines), electronic resources (Internet or compact discs), and other resources. The requirements are the limits to designing or making a product or system.

E. **The process of designing involves presenting some possible solutions in visual form and then selecting the best solution(s) from many.** Sketches or drawings should be used because they provide visual records of the possible solutions. Complete and accurate records of the work should be kept.

F. **Test and evaluate the solutions for the design problem.** Use criteria identified in the requirements for evaluating the solutions. After selecting a solution, build it to show the design idea. Also observe safety when using tools and materials. Through this process, one will gain experience with various types of materials — from measuring, marking, cutting, and shaping to assembling and combining.

G. **Improve the design solutions.** Repeating steps in the design process may be necessary to optimize the design before communicating the results to others. If the solutions are not optimal at this point, the students will return to the design process.

GRADES
6-8

In grades 6-8, students are restless, energetic learners who enjoy active, hands-on experiences. The benchmarks at this level call for students to apply a design process that will enable them to develop their ideas in greater detail and to create their design solutions on a larger, more complex scale. They need to recognize that multiple ideas may solve a problem. Before designing a solution, students must specify goals that will establish the desired results for the problem. These goals will then be used to guide the design process, and ultimately, they will be used to evaluate the final product or system.

After establishing the design requirements, students should develop a proposal, which should detail the size, shape, resources, and specifications for making the design. It can include sketches and drawings that incorporate symbols and clarifying notes. Over time, symbols used in the design proposal have become standardized and have come to represent specific components.

At the middle-school level, models are formally introduced. Using a model is an effective way to simulate what the design will look like. Models can take many forms, such as physical replicas of artifacts, computer programs, conceptual and mathematical modeling, and simulated products. For example, a model of a building is often created by an architect to show clients how it will ultimately look.

After the design proposal has been finalized and the model has been created, it is important to perform tests and evaluate the results as they relate to the pre-established criteria and constraints. This testing and evaluating allows students to refine the design proposal before it becomes a reality. Once they begin the process of making their designs, students should continue to evaluate their ideas in hopes that the final solution will be the best one possible.

Students should actually build the solution(s) as a final activity. If any problems with the proposed solution surface, some of the steps in the design process can be repeated, not necessarily in the same order, to obtain the optimum solution. It is important for students to document procedures and results as they go through each step of the design process. They should communicate their successes, as well as their disappointments. Through this process, students will gain valuable insights from one another. Various techniques for documentation include design portfolios, sketches, journals, schematics, and World Wide Web pages.

As part of learning how to apply design processes, students in grades 6-8 should be able to

H. **Apply a design process to solve problems in and beyond the laboratory-classroom.** Perform research, then analyze and synthesize the resulting information gathered through the design process. Identify and select a need, want, or problem to solve, which could result in a solution that could lead to an invention (an original solution) or an innovation (a modification of an existing solution). Identify goals of the problem to be solved. These goals specify what the desired result should be.

I. **Specify criteria and constraints for the design.** Examples of criteria include function, size, and materials, while examples of constraints are costs, time, and user requirements. Explore various processes and resources and select and use the most appropriate ones. These processes and resources should be based on the criteria and constraints that were previously identified and specified.

J. **Make two-dimensional and three-dimensional representations of the designed solution.** Two-dimensional examples include sketches, drawings, and computer-assisted designs (CAD). A model can take many forms, including graphic, mathematical, and physical.

K. **Test and evaluate the design in relation to pre-established requirements, such as criteria and constraints, and refine as needed.** Testing and evaluation determine if the proposed solution is appropriate for the problem. Based on the results of the tests and evaluation, students should improve the design solution. Problem-solving strategies involve applying prior knowledge, asking questions, and trying ideas.

L. **Make a product or system and document the solution.** Group process skills should be used, such as working with others in a cooperative team approach and engaging in appropriate quality and safety practices. Students should be encouraged to use design portfolios, journals, drawings, sketches, or schematics to document their ideas, processes, and results. There are many additional ways to communicate the results of the design process to others, such as a World Wide Web page or a model of a product or system.

VIGNETTE

The Great Paper Car Race

This vignette exemplifies a group problem-solving activity, in which a paper racecar is created. Criteria are used to evaluate the final solutions. [This example highlights some elements of the Grades 6-8 *Technology Content Standards,* which provide connections with Standards 8, 9, 10, 11, and 18.]

"Ladies and gentlemen, start your engines!" Ms. C told her class. "Your challenge is to design, develop, and produce a racecar that will compete in the second annual Great Paper Car Race at Rolling Hills Middle School. Your car must be designed to roll down an 8-foot ramp into the center of the winner's circle. To make your car, you will be given one piece of 8 1/2" x 11" paper, four wheels, dowel rod axels, thumbtacks, glue, and a limited amount of tape. You will have two days to complete this activity."

Ms. C then divided the class into groups of four to five students, and they began brainstorming various design ideas. Ms. C encouraged them to apply the aerodynamic principles that they had learned in prior lessons and the concepts of force, motion, and the transfer of energy that they had learned in their science class. The groups also had to use problem-solving strategies, critical thinking skills, and teamwork skills as they evaluated each of their ideas and selected the best one.

Each group then used the materials that they were given to build their car. Once the cars were built, each car was timed as it rolled down the ramp. The students were evaluated in three categories: *Teamwork* — Did the team work together? Were they able to produce a completed product? What proportion of planning, designing, and construction did each team member contribute? *Problem solving* — When the team encountered a problem, how did they react? Did the team solve the problem? *Design Solutions* — How well built was the final artifact?

As a result of this activity, the students had an opportunity to experience firsthand how the seemingly abstract concepts and theories they were learning could be applied to concrete, real-life situations. In addition, they gained valuable insights into the value of working together as a team in order to solve a problem.

GRADES

9-12

By the time students graduate from high school, they will be able to apply the design process with a high degree of confidence. They also will be able to work with a whole host of hand tools, various materials, sophisticated equipment, and other resources. They will be capable of synthesizing knowledge and processes and applying them to new and different situations.

In working with the design process at the high school level, students will be applying many of the abilities they learned in earlier grades. They should realize that not all problems can or should be solved. They will also become more adept at identifying requirements in the design process.

The major new skill students develop will be working with prototypes, which can be full-size or scale models, depending on the size of the final product or system. For example, it would be impractical to make a prototype of a skyscraper to full scale. Prototypes and other models should be used to test and evaluate the solutions. Based on these tests, the design solution should be refined if necessary. It is important during this evaluation stage to remember previously discarded solutions in case they are useful later.

Students should be exposed to more sophisticated conceptual, physical, and mathematical models in their late secondary school experiences. Teachers should be careful that the problems their students select are worthy of in-depth research and are challenging enough so that students gain the optimal level of knowledge and skills from the process.

Evaluation becomes even more important at the high school level for testing and judging ideas. Students should be competent at this process and be able to sort through a multitude of possible solutions, test and evaluate them, and determine the best ones for the problem at hand.

Based on the results of the testing and evaluation, the process of building the designed product or system can proceed and may result in one or many items. Students need to see how the building phase goes and then to use that information in evaluating the design. Students also should be challenged to consider economic factors when designing their product or system. In addition, they should consider the sustainability and disposability of the resources in the final product.

As part of learning how to apply design processes, students in grades 9-12 should be able to

M. **Identify the design problem to solve and decide whether or not to address it.** It is important to determine whether the design problem is worthy of being addressed or solved. If the problem is worthy of being solved, students should research, investigate, and generate ideas for the design. Brainstorming is an excellent technique for generating ideas and encouraging creative thinking. Designers often use this technique. Next, synthesize the research and specify the goals of the design. Deductive thinking processes should be used to limit the possible solutions to a few good ones.

N. **Identify criteria and constraints and determine how these will affect the design process.** Identifying criteria and specifying constraints will provide the basis for what the design should be

GRADES
9-12

and what its limits are. Carefully consider concept generation, development, production, marketing, fiscal matters, use, and disposability of a product or system. Test, experiment with, select, and use a variety of resources to optimize the development of the design. If sufficient resources are not available, existing resources could be modified or new ones could be identified. Identify and consider trade-offs among the proposed solutions. Next, plan and select the best possible solution that takes into account the constraints and criteria obtained from research and personal preference. This involves synthesizing various factors, including the constraints, criteria, and information gathered by research.

O. **Refine a design by using prototypes and modeling to ensure quality, efficiency, and productivity of the final product.** Evaluate proposed or existing designs in the real world. Modify the design solution so that it more effectively solves the problem by taking into account the design constraints in order to consider the next step.

P. **Evaluate the design solution using conceptual, physical, and mathematical models at various intervals of the design process in order to check for proper design and to note areas where improvements are needed.** Checking the design solutions against criteria and constraints is central to the evaluation process. Assess previously ignored solutions, perhaps with modifications, as possible choices. When previously favored solutions are discarded, they may be still appropriate for consideration later in the design process.

Q. **Develop and produce a product or system using a design process.** Sometimes items can be produced in single quantity, while others can be made in batches or volume production. Quality control ensures that the product is of high enough quality to be sold.

R. **Evaluate final solutions and communicate observation, processes, and results of the entire design process, using verbal, graphic, quantitative, virtual, and written means, in addition to three-dimensional models.** The final results should be compared to the original goals, criteria, and constraints.

VIGNETTE

The America's Cup Challenge

In this vignette, students participate in an America's Cup competition of their own by designing and constructing a wind-powered monohull vessel. The winner receives the "Cup." [This example highlights some elements of the Grades 9-12 *Technology Content Standards* that provides connections with Standards 8, 9, 10, 11, and 18.]

The Challenge

Design a wind-powered monohull vessel to travel a predetermined distance in the least amount of time.

Size Constraints

Maximum overall length — 16 inches. Maximum hull width (beam) — 8 inches. Maximum overall height — 24 inches. Maximum boom length — 16 inches.

Time Constraints

Items due Week One: 1) Nautical terminology/definitions; 2) Full-size deck plan; 3) Full-size profiles; 4) Determination of hull material.

Items due Week Two: 1) Rigging design; 2) Determination of sail material.

Items due Week Three: 1) Additional nautical terminology.

Items due Week Four: 1) Vessel ready for trials; 2) Written and oral report.

Construction Constraints

Dugouts and solid hull vessels are not acceptable. The sail width must not exceed the length of the hull.

Organizational Criteria

1. The racing team will consist of a captain and two crew members.
2. The vessel will represent a country other than the United States.
3. The vessel must be named (lettered on the vessel) in the language of the country represented.
4. The vessel must fly the flag of the country represented.
5. The written and oral reports must be organized in accordance with the assigned instructions.
6. All construction must take place within the confines of the school.

Evaluation

Individual and group input will be used to assess the success of the group's design solution. In addition, assessment will be based on the results of a race held at a basin or pond located near the school.

Resources

Masts and spars; ballast; sails; molds; adhesives; tools; machines; equipment; methods; research and development information; and testing procedures.

STANDARD

12 Students will develop the abilities to use and maintain technological products and systems.

Everyone uses technological products and systems — cars, televisions, computers, and household appliances — but not everyone uses them well, safely, or in the most efficient and effective manner. Much of the problem lies with the rapid pace of technological change. New technologies appear so frequently that it can be difficult to become comfortable with one before the next has taken its place. As a result, people are driving cars whose minor failings they are incapable of diagnosing; they are working with computers most of whose capabilities are a complete mystery; and they are staring at a flashing "12:00" on their VCRs.

A technologically literate person does not necessarily know how to use every technology safely and effectively, but, when necessary, he or she is able to learn to use a particular technological product or system and is comfortable doing so. At this level, students should be exposed to a variety of technologies, including the newest, and they should be taught the proper learning tools for mastering them, beginning with reading the instructions and owner's manual. Students should learn to select appropriate technologies for a given situation. They should be able to analyze malfunctions and come up with appropriate responses. They should recognize that in some cases the proper use of a technological product is simply not to use it at all.

Appropriate maintenance of a product or system is crucial to keeping it in correct working order, and, when malfunctions do happen, appropriate repair is necessary. Troubleshooting, testing, and diagnosing are important processes in maintaining and repairing a product or system.

GRADES
K-2

From the earliest grades, students will be exposed to various products and systems, and they will be given opportunities to use them correctly and to learn what happens when they are used improperly. For example, students could learn how to use a clock to tell time, how to use a telephone correctly, and how to use basic hand tools properly. The students should be encouraged to investigate each item, perhaps by taking it apart or by comparing it to similar items to discover how it works, its use, and its purpose.

Young students are interested in everything they see around them and are asking questions about how things work, why things are a certain way, and how things came about. Students should be encouraged to find answers to their questions using various tools available to them. Children should be encouraged to follow directions — a type of communication that offers guidance on how to use a tool or product correctly. Directions can be written, verbal, or step-by-step illustrations.

Employing products and systems often requires students to use common tools, such as staplers, screwdrivers, rulers, scissors, and clamps. Although many students will have used tools before, they may not know how to use them correctly. Through formal and informal learning activities and guided discussions, students will learn the best and safest way to use tools.

Symbols are also important in the communication process. Students should recognize that symbols are all around them, from logos representing their favorite sports teams to warning signs on roads. These symbols communicate information and directions in an efficient manner, and they allow children to "get the message" without using a lot of words.

As part of learning how to use and maintain technological products and systems, students in grades K-2 should be able to

A. **Discover how things work.** This can be done by carefully taking something apart (while making a sketch of how the parts fit together) and then putting it back together. The ability to observe, analyze, and document is vital to successfully accomplish this task.

B. **Use hand tools correctly and safely and be able to name them correctly.** Tools have always provided a means for humans to extend their capabilities. Simple tools such as scissors, needles and thread, staplers, hammers, and rulers are examples of devices that everyone needs to know how to use.

C. **Recognize and use everyday symbols.** Symbols are used as a means of communication in the technological world. Examples include road signs, symbols for disabled people, and icons on a computer screen.

GRADES
3-5

Building upon knowledge from grades K-2, students will learn more about how to use products and systems and what should be done if they are not working properly. Reading and following instructions from a users' manual is an important first step in assembling and using products and systems correctly. Students must learn to follow step-by-step directions. Teaching this ability will require a significant period of time with many exposures to directions that are both well and poorly done. Students will need to learn the appropriate questions to ask when directions are not available or are not clear. At this level, students will practice taking a product or system apart and reassembling it in order to learn how things fit together and work. The knowledge gained in such exercises will help them when they use and troubleshoot other products and systems. For example, students could take a toy car apart to see how the gears and the steering system work. They can then apply their new knowledge to investigate why a toy car does not roll or change directions properly.

Given many opportunities to use tools, students should become proficient in selecting the best one for a given task. Students must also be taught to keep safety foremost in their minds when they are using tools. Tools that help students access, organize, and evaluate information should receive special attention. Such tools should include newer resources like computers, CD-ROMS, or the Internet, in addition to the more traditional print sources.

In addition, students should understand and be able to use various symbols in different settings. These symbols could include signs in the community and icons on computers. In classroom activities, students may be challenged to create new symbols that could be used in the home, school, or community to begin to understand the need for symbols and how they aid in communicating key ideas quickly.

As part of learning how to use and maintain technological products and systems, students in grades 3-5 should be able to

D. **Follow step-by-step directions to assemble a product.** These directions could come from a paper or booklet that describes how to put something together or how to solve a problem.

E. **Select and safely use tools, products, and systems for specific tasks.** Tools should be selected based on their function (what they are designed to do), ease of use, and availability.

F. **Use computers to access and organize information.** This could be done with software on the computer (for example, an encyclopedia on a CD), as well as on the Internet.

G. **Use common symbols, such as numbers and words, to communicate key ideas.** Most of these symbols are found in everyday life, such as the alphabet, numbers, punctuation marks, or commercial logos.

VIGNETTE

Take It Apart

This vignette deals with an assembly line operation to find out how pens can be disassembled then reassembled in an efficient manner. Also more intricate items can be brought into the classroom for students to take apart and put back together. This process teaches how parts fit together and how they work. This vignette highlights some elements of the Grades 3-5 *Technology Content Standards* which provide connections to Standards 8, 9, 10, 11, 12, and 19.]

Using a lesson from *Mission 21: Launching Technology Across the Curriculum Series*, Ms. W decided to introduce technology to her third graders. Primary objectives of this lesson were to "increase students' sequencing skills and broaden her students' concepts of what technology is by exploring its influence in commonly used [items]" (Dunlop, Croft, & Brusic, 1992, p. 9). Ms. W used a design brief from the series titled "Take It Apart." She divided her students into teams and asked them to disassemble retractable ballpoint ink pens.

The students took the pens apart, sketched the components, and labeled each one. This method of documentation helped the students later when they developed their presentations. Members of the teams formed an assembly line to take the pens apart, reassemble them, and discuss how they worked. They used their sketches as a guide in putting the pens back together.

After the pen exercise, the students brought in additional items to explore, most of which were more complex than the ink pen. However, having had the opportunity to explore with the ink pen, the students made the transition with ease. All of them took apart their items, drew the parts, labeled them, and documented how to put the objects back together. They then shared their learning through demonstrations by explaining about the various parts. The students were challenged with their activities, and most wanted to know if they could take something else apart and reassemble it. This type of student-centered activity creates understanding within students and encourages group effort. (Dunlop, Croft, & Brusic, 1992, p. 10).

GRADES

6-8

Students at the middle level will explore, use, and maintain a variety of hand tools and machines, consumer products, and technological systems. They will continue to practice proper safety procedures — wearing protective clothing and eye protection, for example — and to follow directions from manuals and other protocols in order to ensure a safe working environment.

In these grades, students also will learn to use the computer, calculator, and other tools to collect data and analyze information to help them determine whether a system is operating effectively. In addition to correcting problems, students will be taught to be proactive and to establish routine maintenance schedules that will keep their technological products operating efficiently.

In order to manage systems effectively, students need to develop a systems-oriented way of thinking — considering technology in terms of inputs, processes, outputs, and feedback. Students at this grade level will learn to identify types of systems and become familiar with how these systems and subsystems operate. One useful exercise involves assembling several subsystems to create a larger system and then tracking how the various parts interact with and affect the performance of another.

To deal with technologies that have become inefficient or have failed, students will learn the skills of diagnosing, troubleshooting, maintaining, and repairing. They should be able to recognize when a system malfunctions, isolate the problem, test the faulty component or module, determine whether they can correct the problem, and decide if they require outside help. Students should understand and follow maintenance procedures, such as cleaning, oiling, adjusting, and tightening screws, in order to keep products and systems operating properly.

As part of learning how to use and maintain technological products and systems, students in grades 6-8 should be able to

H. **Use information provided in manuals, protocols, or by experienced people to see and understand how things work.** This information is helpful in learning how to use a product and determining if it works properly. In addition, many manuals provide tips on how to troubleshoot a product or system.

I. **Use tools, materials, and machines safely to diagnose, adjust, and repair systems.** For many consumer products, federal and state laws require safety information. Safety procedures should be learned through formal education.

J. **Use computers and calculators in various applications.** Computers can be used to control production systems and to research answers to problems.

K. **Operate and maintain systems in order to achieve a given purpose.** The understanding of how a system works is vital if one is to operate and maintain it successfully. Examples of everyday systems could include the Internet, control systems such as robots, and gating circuits for digital processing of information.

GRADES

9-12

By the time they leave high school, students will be able to use and maintain various types of products and systems — a key element to technological literacy. Some students also will have developed strong personal interests and abilities in technology and will be ready to pursue further education in the field. Students will be able to articulate and communicate their thoughts to others using oral, written, and electronic communication techniques.

Students should be capable of diagnosing, troubleshooting, analyzing, and maintaining systems. These abilities become central to keeping systems in good condition and in working order. Students should be taught the importance of establishing maintenance schedules to prevent breakdowns. For example, they could establish regular schedules to change the air filters in their homes or to change the oil in their cars. If certain products or systems fail, it is important that students be able to diagnose, troubleshoot, and repair them. Much information about how to diagnose and repair a product or system is contained in the product's service manual. Having a clear understanding of the severity of the problem is key to deciding if more experienced help is needed.

As has been stressed at other grade levels, the safe and effective use of tools and machines is important to technological literacy. Students should be given many opportunities to use various tools, such as meters and oscilloscopes, to retrieve, monitor, organize, diagnose, maintain, interpret, and evaluate data and information that can then be used in solving practical problems.

Systems thinking, which combines methodical and analytical thinking, provides the mental skills for students to determine if a system is being used appropriately and correctly. By using such high-level thinking, students will be able to work with inputs, processes, outputs, and feedback to adjust a system.

At the high school level, students should become proficient in using computers and software. They should know about and be able to use advanced computers and peripherals, and they should be able to troubleshoot hardware and software problems. They should also understand the capabilities and limitations of computers. Finally, students should be able to adapt to emerging computing technologies.

As part of learning how to use and maintain technological products and systems, students in grades 9-12 should be able to

L. **Document processes and procedures and communicate them to different audiences using appropriate oral and written techniques.** Examples of such techniques include flow charts, drawings, graphics, symbols, spreadsheets, graphs, time charts, and World Wide Web pages. The audiences can be peers, teachers, local community members, and the global community.

M. **Diagnose a system that is malfunctioning and use tools, materials, machines, and knowledge to repair it.** Various items, such as digital meters or computer utility diagnostic tools, can be used in the maintenance of a system.

GRADES
9-12

N. **Troubleshoot, analyze, and maintain systems to ensure safe and proper function and precision.** Monitoring the operation, adjusting the parts, cleaning, and oiling of a system represent examples of how a product or system can be properly maintained.

O. **Operate systems so that they function in the way they were designed.** These systems may include two-way communication radios, transportation systems that move goods from one place to another, and power systems that convert solar energy to electrical energy. Using safe procedures and following directions is absolutely essential to ensuring an accident-free working environment.

P. **Use computers and calculators to access, retrieve, organize, process, maintain, interpret, and evaluate data and information in order to communicate.** Many resources, such as library books, the Internet, word processing and spreadsheet software, in addition to computer aided design (CAD) software can be used to access information.

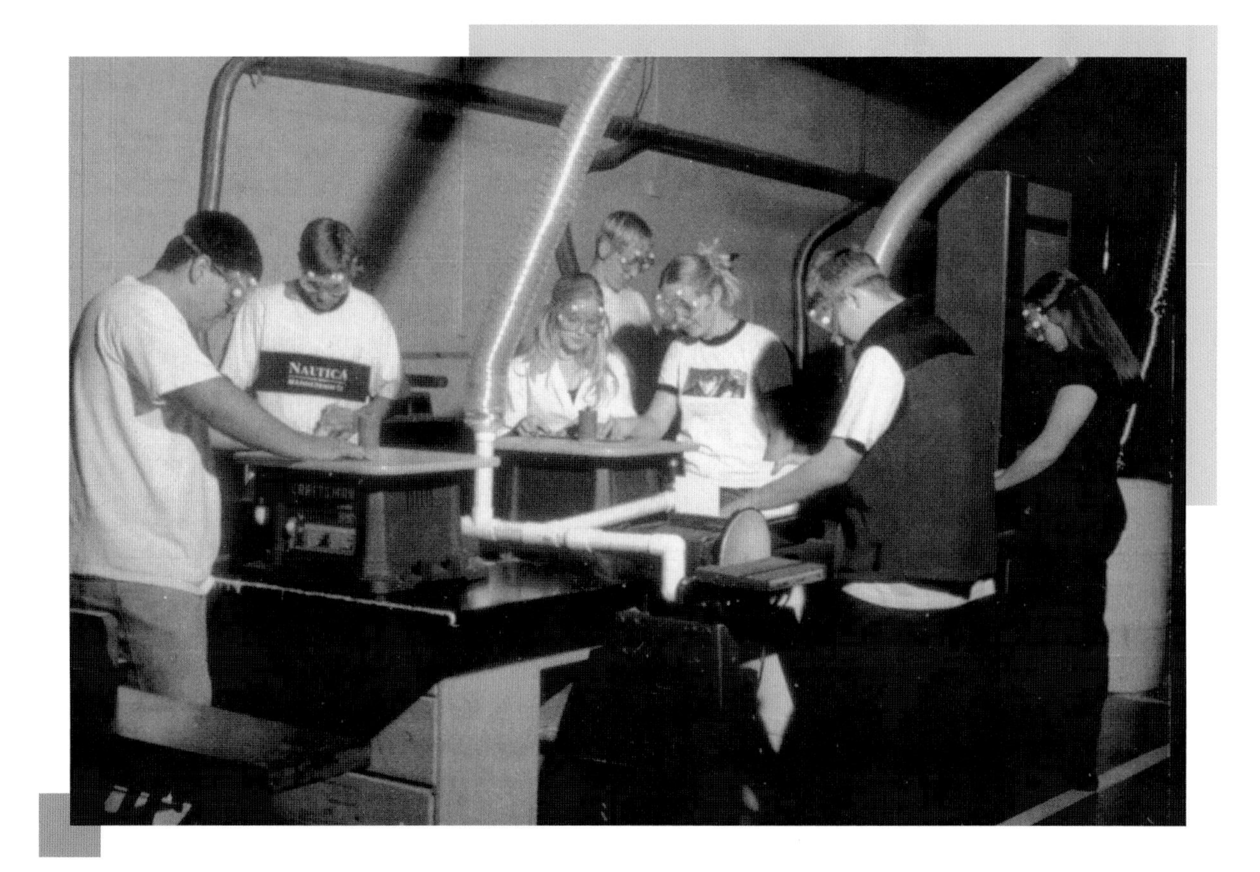

STANDARD

13 Students will develop the abilities to assess the impact of products and systems.

When presented with a particular product or system, the technologically literate person should be able to gather information about it, synthesize this information, analyze trends, and draw conclusions regarding its positive or negative effects. To assess a technology in this way, students should be encouraged to acquire new abilities. They should be able to make forecasts by using a variety of techniques, such as repeated testing, reasoning from past experiences, foreseeing possible consequences, modeling and developing scenarios, and determining benefits and risks. Then, working from these forecasts, they should be able to assess how a product or system will affect individuals, society, and the environment.

This sort of assessment is particularly important today because the human use of technology has become so widespread that it can result in positive or negative consequences, and it is so complex that it can be difficult to predict. Students should realize that technological activities inevitably involve trade-offs, as well as a certain amount of risk. In assessing a technology, they must distinguish between real and imagined risks and recognize that even doing nothing can result in an uncertain outcome.

GRADES

K-2

Technology plays an important part in children's lives. It provides shelter, information, toys, clothing, and food. As current and future consumers, children should be able to determine if the use of a product or system will have positive or negative results. They also should begin comparing products to determine the best value.

Students at this level should collect information about everyday products and systems asking such questions as: Where did the product come from? How was it made? Does the product work well? Can I afford to buy it? Does the product do what it was advertised to do? What are the safety factors to be considered when using the product? How long will the product last? Does the product require additional costs (e.g., paper, film, or batteries)?

Learning to collect information about technology is important in developing an ability to correctly make decisions about its use and in evaluating its effectiveness. The concept of data collection as a means of decisionmaking should be introduced in grades K-2 when students are also studying data collection in science and mathematics. Using easily observable requirements (e.g., numbers, size, texture, weight, and motion), students will identify, categorize, and compare different kinds of technology.

In grades K-2, students should use this collected information to determine whether a product generally produces positive or negative results. For example, students could examine the use of disposable food containers in local fast food restaurants. In this activity, students could list the positive attributes (e.g., makes the lines go faster and gets rid of the need for washing dishes) and negative attributes (e.g., produces a lot of trash) of these containers. Such experiences can help students to begin to develop a critical eye for technology.

As part of learning how to assess the impact of products and systems, students in grades K-2 should be able to

A. **Collect information about everyday products and systems by asking questions.** Examples of some questions are: What are they? Why are they important? Can they be recycled? What is the cost? Where do they fit into everyone's lives? How do they affect daily life?

B. **Determine if the human use of a product or system creates positive or negative results.** Examples could be the positive or negative effects of using televisions, toys, bicycles, games, dolls, and the Internet.

GRADES

3-5

Students in grades 3-5 will have an opportunity to assess technology from personal, family, community, and economic points of view. In assessing technology, students take a step toward becoming self-reliant, independent thinkers. Furthermore, in learning to assess technology, students will develop skills in comparing, contrasting, and classifying collected data, and they will then use that data to make decisions.

Gathering information involves making investigations and observations about the use of technology and then recording the observations in an appropriate manner. Knowing how to gather data requires students to use skills from other areas — notably science skills, like observation, and language-arts skills, such as note taking, outlining, and informative writing.

Students should explore how technology influences individuals, families, communities, and the environment. As students study the significant events that helped to shape their communities, they lay the groundwork for discussing and learning about the development and future use of technological products and systems. They should learn to recognize the trade-offs implicit in any technology and to weigh those trade-offs to determine whether the positive outcomes of a product or system will outweigh its negative consequences.

As part of learning how to assess the impact of products and systems, students in grades 3-5 should be able to

C. **Compare, contrast, and classify collected information in order to identify patterns.** Information, such as cost, function, and warranties, could be collected on certain products, such as toys, food, games, health products, school supplies, and clothes, or on larger systems, such as transportation or communication.

D. **Investigate and assess the influence of a specific technology on the individual, family, community, and environment.** Examples of this could be the family car, microwave oven, clothing, processed food, electric power plants, or passenger airplanes.

E. **Examine the trade-offs of using a product or system and decide when it could be used.** It is important to decide which problems the products or systems are solving and which ones they may create. For example, a question that may be examined is: "Should cars be used?"

VIGNETTE

Clean Up an Oil Spill

This vignette presents a real-life problem of an oil spill. Students are challenged to solve the problem by trying out different activities in the laboratory-classroom. Students should be able to assess the impact of the oil spill. [This vignette highlights some elements of Grades 3-5 *Technology Content Standards,* which provide connections to Standards 3, 5, 6, 10, 13, 16, and 18.]

Ms. G, a fifth-grade teacher, presented her class with a challenge. "On April 5, an oil tanker, *Radical,* sideswiped an iceberg, causing 15 tons of oil to spill into the ocean. You are chosen to join an elite team of scientists to devise a way to clean up the oil spill."

Ms. G then divided the students into groups of four to five members. Each group received a pie pan with one inch of water in it and a small piece of fake fur. Ms. G instructed the students to begin the experiment by placing the fur in the water, and then she used an eyedropper to add a tablespoon of oil to the water in each pan. The students observed what happened and recorded these observations in their journals. The teacher stimulated their thinking by asking questions, such as "Do the oil and water mix? How many layers do you see? What happened to the oil?" The students then estimated the size of the oil spill.

Next, the students received paper towels, cotton balls, toilet tissue, string, coffee filters, and rubber bands. Ms. G then challenged them to use these materials to try to contain the oil in a small area and to clean up the oil spill as much as possible. The students worked together as teams to design a method, and they recorded their observations, their successes, and their not-so-successful attempts in their journals.

Ms. G then led the class in a discussion of the various methods that the students had developed. They discussed the problems that they had encountered in developing a solution and how they had overcome the obstacles. The teacher then asked the students to observe the piece of fur and to describe in their journals how the fake fur had changed and to imagine how oil on an animal's fur might affect its survival. Ms. G then added several drops of detergent to the experimental oil spill, and again she asked the students to write their observations in their journals.

Ms. G concluded the exercise by discussing the importance of petroleum and oil in everyone's life. She also explained the short- and long-term impacts that an oil spill can have on the environment. She shared with them several real-life stories of tanker accidents, the clean-up procedures, the damage to the environment, the community's response to the accident, and the steps taken to design tankers that won't spill oil. As a result of this exercise, the students learned how a technology can have both positive and negative effects on the environment.

GRADES

6-8

At this age, students are able to collect and analyze data and to interpret trends in order to assess what the personal and societal impacts of a particular technology might be. In so doing, students will develop certain important skills. To collect information, for example, students will need to design instruments for gathering data. The collection of data might include conducting surveys, interviewing people, writing letters, or consulting reference sources.

By investigating the effects of technology, students will learn that people react to technology in a variety of ways. For example, students may find that some people favor a proposal for a new shopping mall or golf course in town, while others oppose it. The supporters might be more influenced by the effects on employment, as well as shopping or recreation opportunities, while opponents might be concerned about environmental damage or traffic congestion.

Using and analyzing data will help students begin to appreciate various trends that have occurred in the development and use of technology. Students should learn how to use data to create knowledge. Making sound decisions about technology demands knowledge about current trends. Students will learn what trends are, how they are important in forecasting the future, and how to interpret and make future predictions based on them. After identifying trends, students will learn how to evaluate and monitor the consequences of technological activity. For example, students may use computer simulation software to assess the impacts of technology on a city.

By combining various skills, students can evaluate or assess products and systems to determine if they are useful or not — whether they will achieve the desired outcomes, and whether the positive consequences might outweigh the negative. The ability to analyze a technology critically and objectively is a skill that requires a great deal of time and practice to develop.

As part of learning how to assess the impact of products and systems, students in grades 6-8 should be able to

F. **Design and use instruments to gather data.** Examples of these instruments could be a data-collection instrument for interviews, questionnaires to be mailed, or computer-based forms on the World Wide Web. Assessment tools also could include devices designed to conduct tests on such things as water quality, air purity, and ground pollution.

G. **Use data collected to analyze and interpret trends in order to identify the positive or negative effects of a technology.** Technologically literate citizens are able to fulfill their personal and social responsibility to assess technology.

H. **Identify trends and monitor potential consequences of technological development.** Trends are patterns of technological activities that show a tendency or take a general direction. Trends are used to provide direction in deciding if a product or system should be used.

I. **Interpret and evaluate the accuracy of the information obtained and determine if it is useful.** Developing specific criteria for what is useful is important in making these judgments. Sometimes determining accuracy is easy — taking information from physical measuring devices like a water-purity tester, for example. At other times, accuracy is more difficult to determine, as when assessments are based on public opinion, which can differ greatly from group to group and from time to time.

GRADES

9-12

By learning how to assess technology, students will become better citizens in the future and, as a result, they will be able to make wiser decisions in an increasingly complex technological world. Students should know that the advantages of technology far outweigh the disadvantages. If it were not for the development of technology, humans would be living in a much more primitive world today.

Collecting and synthesizing data is invaluable for making informed technological decisions. For example, people who are interested in buying a product or system may design a forecasting instrument and collect data in order to assess a technology's overall efficiency and intended function. In grades 9-12, students will learn to synthesize data and to use their syntheses to draw conclusions about the use of technology and the effects of its use on individuals, society, and the environment.

It is important for students to be able to use trend analysis to judge trends and to determine what is important in light of other current events. For example, students could research various climate forecast models and project what could occur if the earth's polar regions warmed by 2°C or 4°C. They then could analyze a plan to address global warming and assess its potential solution.

Once information has been accumulated, synthesized, and used for forecasting, the final step in assessing a product or system is deciding whether using it is appropriate. In making such a decision, students should come to understand the benefits and risks, costs, the limits and potential, and the positive and negative impacts of technological developments.

As part of learning how to assess the impact of products and systems, students in grades 9-12 should be able to

J. **Collect information and evaluate its quality.** This may include using such methods as comparing and contrasting sources, examining relevancy, and investigating the background of experts.

K. **Synthesize data, analyze trends, and draw conclusions regarding the effect of technology on the individual, society, and the environment.** Deductive thinking and synthesis techniques can assist in this process. Students should take into account historical events, global trends, and economic factors, and they should evaluate and consider how to manage the risks incurred by technological development.

L. **Use assessment techniques, such as trend analysis and experimentation to make decisions about the future development of technology.** Assessment is an evaluation technique involving iterative steps and procedures that requires analyzing trade-offs, estimating risks, and choosing a best course of action. The assessment of a product or system can prove that it is dangerous, but it cannot prove that it is safe.

M. **Design forecasting techniques to evaluate the results of altering natural systems.** These techniques should include testing and assessment. These natural systems could be lakes (building homes around the shore), rain forests (cutting them down for the wood), or land (strip mining for coal).

7 The Designed World

STANDARDS

7 The Design World

Humans live in three worlds: the natural world, the social world, and the designed world.

The natural world consists of plants and animals, earth, air, water, and fire — things that would exist without human intervention or invention. The social world includes customs, cultures, political systems, legal systems, economies, religions, and the various other mores that humans have devised to govern their interactions and relationships with one another. The designed world consists of all the modifications that humans have made to the natural world to satisfy their own needs and wants. As its name implies, the designed world is the product of a design process, which provides ways to turn resources — materials, tools and machines, people, information, energy, capital, and time — into products and systems.

In studying the designed world, it is useful to create a taxonomy, or classification system, that divides technology into smaller pieces that can be explored individually. The following taxonomy represents important areas that can be studied in K-12 programs. Each area of technology covered in this chapter contains a set of characteristics that defines it and distinguishes it from the others. As the areas of technology change over time, so too will the taxonomy. These areas are not mutually exclusive — there is some natural overlap between them — but dividing them up in this way makes it easier to study the many and varied technologies that humankind has invented. It is possible to create many different taxonomies of the designed world, but for the purposes of *Technology Content Standards*, the designed world is divided into seven standards.

STANDARD

Students will develop an understanding of and be able to select and use medical technologies.

People in today's health-oriented society spend more time and money than at any other time in history searching to live longer and more productive lives. The use of technology has made numerous contributions to medicine over the years. Scientific and technological breakthroughs are at the core of most diagnostic and treatment practices. For example, major operations used to require long hours of surgery followed by an extensive hospital stay. Today, with lasers, new drugs, and updated medical procedures, long hours in the hospital operating room have been decreased to outpatient procedures in a doctor's office, and recuperation time has been reduced from weeks to days.

Medical miracles are cited often in the news — the reattachment of a limb or the saving of a life through a new medical procedure made possible by a new device or system. New ways of studying how the human body functions or reacts to change are being introduced at rapid rates. Devices and systems are being designed to check, evaluate, and operate with computerized and electronic controls in order to extend human capabilities and help improve human health.

The development of good nutrition and preventive medicine has played a key role in helping individuals live better lives. Medical advances, such as vaccines and genetically engineered drugs, are developed to help healthcare providers do their work more efficiently and effectively, thus improving the delivery of medical care. Today, technologies, such as telemedicine (the use of telecommunications technologies to deliver healthcare), are being designed and developed to provide easier access to medical expertise, to integrate geographically dispersed services, to improve the quality of care, and to gain maximum productivity from expensive medical and technical resources.

With the increased use of technologies in the medical industry, it is important to consider the consequences that accompany them. Technologies, such as pharmaceuticals and life support systems, have helped protect and improve human health. However, the use of these products and systems have raised questions, such as the length of time a person chooses to remain on a life-support system and chooses to have access to life-saving procedures.

In 1900 an American's life expectancy was 47 years; today, it exceeds 76 years. Worldwide, human life expectancy has been prolonged through the development of sanitation practices, vaccines, waste-disposal systems, and other technologies. This increase in life expectancy is a central reason for the worldwide population explosion. The issues surrounding the use of many technologies often conflict with each other or with the opinions and ethics of those affected by its use. Knowledge based on accurate information is therefore essential for making sound decisions.

GRADES

K-2

When children enter kindergarten, many will already know that staying clean and healthy habits are important in preventing sickness. They also will know that using certain products and systems help them stay safe and healthy. The K-2 classroom should provide students with the opportunity to learn about habits that promote healthy living, along with the technologies that make it possible.

Oral vaccines, shots, and medicines are used to help prevent diseases or slow their progression. Students should have opportunities to explore how science and technology are used to help promote good health. They may discuss how a germ may cause an illness and then discuss how a product is designed to help prevent the illness or provide a means for a person to get better.

Students are aware of various tools used to examine their bodies when they visit a doctor, dentist, or optometrist. They know that tools, such as a thermometers, scales, dental tools, and optometrist lenses are used to gather information. They should understand how various technologies are specially developed to provide unobtrusive ways to learn more about their health. For example, they could explore how chewable tablets are designed to reveal plaque build-up by coloring their teeth. They might study a stethoscope by taking it apart and examining how sound is used to provide clues about the health of their hearts and lungs.

In order to select, use, and understand medical technologies, students in grades K-2 should learn that

A. Vaccinations protect people from getting certain diseases. Vaccinations help build protection to disease and are often administered early in life. Some immunizations are given, over a period of several months. Vaccinations and shots have led to improved health and life expectancy.

B. Medicine helps people who are sick to get better. Some medicines require a long period of time before they become effective and require repeated doses. Others work in a short period of time and are needing to be used only when needed.

C. There are many products designed specifically to help people take care of themselves. Everyday products, such as toothbrushes, hairbrushes, and soap are used to promote healthy living. Doctors, dentists, optometrists, and other health professionals use many technological tools to gather medical information about people's health.

GRADES

3-5

Specialized products and systems can be used to collect information about many different things that can affect people's health and safety. Doctors use such devices as stethoscopes, x-ray machines, and thermometers to gain clues about why a person may be sick and to help determine what medical attention may be needed.

Students in grades 3-5 should be aware of these and other products and systems that play an important role in keeping them healthy and safe. News reports in the fall inform people about health issues regarding upcoming flu seasons. With increased scientific and technological knowledge about how germs or bacteria work to cause illnesses, more advanced inoculations are designed and produced. Students should continue to make connections on how science and technology work together to promote and improve health.

People who have, through injury or illness, lost body parts or functions can often be helped through the use of medical technologies. Hearing aids can compensate for a loss of hearing, for instance, and artificial limbs help people who have lost arms or legs to lead more normal lives. Students should discuss these and other ways that technologies have been used to help those with disabilities or medical conditions.

Since more and more people spend time in closed environments, students need to understand about the effects of poor indoor air quality and how it can decrease performance and cause illness. Students could visit a hospital or factory or even tour their school in order to learn about the different technological means that have been developed to promote a healthy living and working environment.

In order to select, use, and understand medical technologies, students in grades 3-5 should learn that

D. **Vaccines are designed to prevent diseases from developing and spreading; medicines are designed to relieve symptoms and stop diseases from developing.** Vaccines for such illnesses as polio, tetanus, and mumps are used in the maintenance of good health, while medicines, such as those for the common cold, the flu, or pneumonia are used to help ease an illness and restore good health.

E. **Technological advances have made it possible to create new devices, to repair or replace certain parts of the body, and to provide a means for mobility.** Products such as artificial limbs, wheelchairs, or crutches change to take advantage of new technologies and to improve upon previous designs.

F. **Many tools and devices have been designed to help provide clues about health and to provide a safe environment.** Tools, such as thermometers, blood pressure machines, and heart monitors help determine if people are well and provide other health clues. For example, a heart monitor measures a person's heart rate. Many tools have been designed to diagnose what is happening in the human body. Self-testing kits for glucose, sugar, and pH levels, and kits to determine the levels of protein or vitamins in the body are examples of such tools. This information may help determine if a person's health is stable, or if he or she is developing an illness.

VIGNETTE

A Pharmacy Connection

This example uses a visit to a local pharmacy to encourage students to develop and put to use their understanding of how the design of a vaccine or medicine relates to the process of design and how vaccines and medicine are related to various technological devices. [This example highlights some elements of Grades 3-5 *Technology Content Standards* 1, 6, 8, 11, and 14.]

The students in Ms. B's fifth grade class visited their local drug store to learn more about various medicines in general. They focused particularly on individualized kits designed to help people learn more about their bodies.

Ms. A, the local pharmacist, showed the students how people use various kits to check their bodies' pH and glucose levels, as well as their protein and enzyme levels. The students were particularly interested in the saliva testing kits used to determine sugar levels. In addition, the students noticed all the different devices available for checking their temperature — from the traditional thermometer to the new strips used on the forehead. Ms. A also demonstrated how the electronic ear thermometer worked.

When they were shown the numerous drugs kept in a pharmacy, the students were amazed. They asked Ms. A how she was able to keep track of all the information about the many drugs and customers. She explained that thousands of records were kept in large filing cabinets before they had computers. Ms. A further explained that computers not only enable pharmacists to link customer information with doctors' orders, they also help in delivering advice about how to use medicines safely.

After the students returned to their classroom, Ms. B asked them to investigate more about the development of various medicines and vaccines, in addition to finding out about the tools used in their development. She asked the students to refer to lessons they had studied on design and to consider what processes of design may have been used in the development of a medicine or vaccine. A few students used the Internet to check information, and others referred to several books about medical technologies. After the students had written down their information, Ms. B provided them with an opportunity to share their findings. The students reported that in order for many vaccines to work, physicians send things into the body that tell them how the body works. They are then able to determine if a vaccine is functioning properly. Many students commented on how similar the design and use of a vaccine is to the design and use of a product or system.

GRADES
6-8

In grades 6-8, students are interested in themselves and their own bodies. Learning about technologies designed to protect and keep them healthy is a natural extension of that interest. Students can research and discuss the various technologies that have enabled people to live longer, more productive lives. They could also discuss personal experiences in which a technology has helped them with a medical situation, such as getting glasses or braces.

Students should have opportunities to explore and discuss recent medical developments to learn what types of technologies are used to improve medical care in different environments. For example, the use of many new technologies, such as lasers for surgeries, electronic devices for monitoring and evaluating health, and modified treatment procedures, have helped to increase the well-being of individuals. Likewise, students should consider how some people view the health care system as being dominated by the use of technology. They should learn about how the uses of medical technologies have helped advance medicine from inactive care that was primarily diagnostic in nature to a field dedicated to the aggressive treatment and prevention of disease. In learning how different medical technology devices work, students could design and build models that would demonstrate how they are used.

Genetic engineering is the manipulation or modification of an organism's genes to increase production or remove a genetic defect. Genetic engineering is often discussed in contradictory terms. Although using it is viewed as a powerful means for curing genetic diseases, it is criticized for the potential harm that may come to people and the environment. Students should, therefore, have opportunities to assess different aspects of genetic engineering. For example, they could investigate what technologies are used in genetic engineering, how genetic engineering is used in the health care field, and how gene therapy may affect medical costs and care.

In order to select, use, and understand medical technologies, students in grades 6-8 should learn that

G. **Advances and innovations in medical technologies are used to improve healthcare.** A super-quick, ultra-low radiation digital X-ray machine, originally developed to detect diamond smugglers has been adapted to save lives. Patients undergo a full-body X-ray, which can locate bullets and pinpoint fractures in seconds.

H. **Sanitation processes used in the disposal of medical products help to protect people from harmful organisms and disease, and shape the ethics of medical safety.** Proper handling and management of hazardous materials, such as medicines, clothing, and instruments, help to protect people from unnecessary harm and increase risk-free environments.

I. **The vaccines developed for use in immunization require specialized technologies to support environments in which a sufficient amount of vaccines are produced.** Immunization is the process of systematically vaccinating people through a series of shots to prevent disease. The technological system designed to create the proper environment in which a vaccine

GRADES
6-8

may be cultured is paramount to the success of the large quantity of the vaccine needed for immunization. Increasing the production of a vaccine requires understanding how an organism is tailored to produce a vaccine and how a vaccine works, in addition to addressing the quantity needed for all concerned and providing enough materials for proper production of the vaccine.

J. **Genetic engineering involves modifying the structure of DNA to produce novel genetic make-ups.** Genetic engineering is done in a laboratory using reagents and other tools that allow researchers to make controlled changes in genetic information and structure. A practical example in the area of molecular pharmaceutical industry involves the process to remove the human insulin gene from human cells and to move it into bacterial cells (*E. coli*) with other genetic signals that instruct the bacteria to make human insulin. Large amounts of human insulin can be produced by this recombinant DNA method. The human insulin (Humulin) has been found to be superior to that derived from the pancreas of pigs (porcine insulin) because patients using pig insulin can develop allergies that can compromise the effectiveness of diabetic treatment.

GRADES
9-12

At this level, the technologies used for health and medicine should be critically researched and discussed, including the global concerns regarding the environment and the ethical concerns about altering life. Students should gain the ability to debate such questions as: How do people know when a medical technology is beneficial? At what point should people be involved in the testing of a medical invention and innovation? To what degree are designers responsible for the safety of their products or systems? How will certain products and systems affect the current and future environment?

Students should have opportunities to identify emerging health and medical technologies, such as genetic engineering, noninvasive surgeries, arthroscopies, and nuclear magnetic resonance, by using trends, research, and forecasting techniques. For example, students could study and learn how a laser works by making, testing, and evaluating a model and then relating its adaptation to use in many surgical procedures. They should communicate their findings to a wide variety of audiences, including peers, family, and the community, in order to explain their viewpoints on how products and systems can be used to promote safe and healthy living.

Advances in medical technology have helped to improve human health by reducing the instances of serious diseases, such as polio and smallpox. Yet there is still a great need for continued improvements and more innovations. Students should investigate both the benefits of the advances in medicine made through the use of technology and the associated costs. In addition, students should be aware of how technology is being used regarding such topics as population control, gene mapping, and the medical effects of pesticide usage. Similarly, students should examine how computers in the healthcare system play an integral role keeping track of patients' diagnostic information, medicines, results of procedures, and in analyzing data in order to help clinicians do their work more efficiently and effectively.

In order to select, use, and understand medical technologies, students in grades 9-12 should learn that

K. **Medical technologies include prevention and rehabilitation, vaccines and pharmaceuticals, medical and surgical procedures, genetic engineering, and the systems within which health is protected and maintained.** For example, the development of vaccines and drugs, such as the polio vaccine, penicillin, and chemotherapy, has helped to eradicate or cause remission of many serious illnesses. The development of diagnostic tools, such as the x-ray machine, computerized tomography (CT) scan, and lasers, allows for less invasive interior views of the body than surgery. The use of specially designed equipment can help provide rehabilitation to disabled persons. Using a wheelchair and other specially designed equipment, a paraplegic person can play basketball; dialysis maintains health for those with no kidneys; and laser eye shaping helps eliminate the need for glasses or contact lenses. Many technologies designed for health, medicine, and safety are specialized and can be expensive to maintain.

GRADES
9-12

L. **Telemedicine reflects the convergence of technological advances in a number of fields, including medicine, telecommunications, virtual presence, computer engineering, informatics, artificial intelligence, robotics, materials science, and perceptual psychology.** Telemedicine is designed for emergency situations, rural health care, forensic medicine, and monitoring chronic conditions. Telemedicine represents a significant change in the delivery of medical care by increasing the number of doctors who can diagnose illness and offer treatment in unsafe and remote areas via computer or videoconference. For example, when a scientist in Antarctica discovered a potentially cancerous lump and could not fly out for medical care, equipment was airlifted to the location, and doctors, located in the United States, were able to use the medical equipment and communication devices in the determination of a treatment.

M. **The sciences of biochemistry and molecular biology have made it possible to manipulate the genetic information found in living creatures.** Recombinant DNA technology, in the form of applied molecular research, has resulted in methods for screening and diagnosis of disease states and disease predisposition (molecular diagnostics). The potential for misuse of this information should compel society to establish ethical mandates for regulating the incidence of testing and the uses of test results.

STANDARD

15 Students will develop an understanding of and be able to select and use agricultural and related biotechnologies.

About 14,000 years ago, the Agricultural Revolution transformed society by allowing humans for the first time to produce more food than they needed. The development of a variety of agricultural tools and practices, such as the plow and irrigation, improved productivity and made it possible for fewer people to feed all of a society, thereby freeing up some of the society for other tasks. Further advances in agricultural technology since that time have continued the pattern, so that now only about one out of a hundred people working on a farm is enough to provide food for everyone in the United States.

Agriculture is the growing of plants and animals for food, fiber, fuel, chemical or other useful products. Many technological processes and systems are used in agriculture. One example of a simple process is the saving of seeds from the end of one growing season to plant at the beginning of the next. Another is the use of fertilization and weed control. Breeding plants and animals in order to produce offspring with desired traits is yet another example of agricultural technology. Also, of course, there is a long line of agricultural tools and machinery — from pointed sticks used to scratch a line in the soil for planting seeds to today's most advanced combines or milking machines. Technology has not only improved the yield and quality of food, but it has also made it possible for farmers to adjust to changing circumstances in the environment, such as weather-related changes, water shortages and floods, and overused soil.

Biotechnology applications, both classical and modern, have long been a driving force in the field of agriculture. Biotechnology is defined as "any technique that uses living organisms, or parts of organisms, to make or modify products, improve plants or animals, or to develop microorganisms for specific purposes" (OTA, 1988/1991, FCCSET, 1992/1993). It encompasses a broad spectrum of purposes from changing the form of food and improving health to disposing of waste or using DNA to store data. Living organisms involved in biotechnology can be microorganisms, plants, and animals, as well as their parts (i.e., enzymes and proteins). Although it has a modern ring to it, biotechnology has been around for at least 8,000 years. Around 6000 BC, for example, Babylonians used yeast to brew beer, and about 4000 BC Egyptians learned how to use yeast in making bread.

In recent years, the use of biotechnology has become vastly more important as scientists have become more proficient at manipulating cells and living tissue. They have learned to read the genetic codes of living organisms and to manipulate their genetic instructions. The use of biotechnology is opening the door to improving the fight against human and animal diseases, promoting human health, fighting hunger by increasing crop yields through resisting plant diseases, and helping the environment by reducing pesticide use, but this is just the beginning. Experts predict that an explosion of new products and services will result from biotechnology advances over the next century.

More than other types of technology, biotechnology tends to raise ethical and social issues. How safe are bio-engineered crops? Should society allow the use of bioengineering methods? If society is to answer such questions responsibly, its members must have a basic understanding of biotechnology and the resulting products and systems involved.

All of the technologies designed and used for agricultural and related biotechnology products and systems have an effect on the environment. The term "artificial ecosystem" is used in this standard in its broadest sense, to include the design and manipulation of nature to create such artificial ecosystems as farms, ponds, gardens, and human-made forests. These artificial ecosystems are designed to provide food, fiber, fuel, chemicals, and other goods. A clear understanding is needed of the limitations of the technologies used to create the artificial ecosystems and how to use and manage them effectively in order to sustain the earth's natural resources.

GRADES

K-2

Students who live on farms or have a garden in their yards will be familiar with some of the products and systems used to grow plants and raise animals, while children who live in cities may have only a vague understanding of the technologies and processes. As a result of experiences in grades K-2, all students will have a basic understanding of how the food they eat and how the clothes they wear are produced.

Living things depend on air, energy (sun), food, and water. If any one of these basic elements is missing, plants and animals will not survive. Students will discover how all of these elements work together with technological products to form a system that enhances the growth of plants and food. For example, they could plant several pots of seeds using various types of soils. The different pots could be watered with different amounts of nutrients to demonstrate to students how agriculture relies on the use of supplements to enrich the soil in various conditions. Some of the pots could be placed in the direct sunlight, while others could be placed in the dark. The students could then observe what happened to the different pots.

Preparing a "biology-in-the-bottle" ecosystem will help students learn about using the design process, working with tools, and conserving water. Using a container that provides a closed environment, students decide what they want to grow and then research how much water and sunlight the plants need. The students could then prepare the soil, plant the seeds, and place them in a container. This activity will enable students to begin to understand how plants are grown and what techniques are needed to care for them in an artificial ecosystem.

In order to select, use, and understand agricultural and related biotechnologies, students in grades K-2 should learn that

A. **The use of technologies in agriculture makes it possible for food to be available year round and to conserve resources.** The processes of planting, growing, maintaining, harvesting, and preserving are important in providing food. Conservation of water requires using it wisely in homes, yards, gardens, and farmlands.

B. **There are many different tools necessary to control and make up the parts of an ecosystem.** An ecosystem is the collection of organisms, such as plants and animals, in a shared physical environment. Understanding how plants, animals, and their wastes interact with their environment is important in order to know how to use them as natural devices for scrubbing or cleaning the environment. For example, trees and grasses are used to help remove carbon dioxide from the air and generate oxygen, while lakes, rivers, and marshlands help to maintain and conserve water for all to use.

GRADES

3-5

Plants and animals, just like students' own bodies, require scheduled care. By applying what they have learned in their science lessons about how things grow, students can research how the food that they eat is produced. Because water is essential for living things, students should explore different techniques for moving water to its desired location, how to conserve it, and how to keep it from becoming polluted. They also should investigate what transportation systems are used to move agricultural products from place to place.

At this level, students also should design, build, and assess artificial ecosystems for different organisms. Students should research how much food, space, and sunlight organisms need in a given ecosystem, and how plants give off oxygen that in turn is needed by the animals. They could apply this sort of information by raising a hamster in their classroom. In designing the hamster habitat, students should be attentive to how their system provides clean water for drinking and the types of materials used for bedding and collecting waste. Students could observe the composting process developing in the material of their designed ecosystem. They might design and make a wildlife habitat by landscaping a small area in their schoolyard to attract birds, butterflies, beneficial insects, or small animals. In conjunction with this activity, students could read and write about wildlife habitats and their experiences in designing and making one. Through these various activities, students will gain a better understanding of various technological processes used in agriculture, including propagating, growing, maintaining, evaluating, and harvesting.

In order to select, use, and understand agricultural and related biotechnologies, students in grades 3-5 should learn that

C. **Artificial ecosystems are human-made environments that are designed to function as a unit and are comprised of humans, plants, and animals.** A farm or a garden pond is an example of an artificial ecosystem designed to use all parts to their fullest potential. A farmer uses the plants, animals, and land to work together for optimum production. A garden pond is designed to use plants to provide food and shelter for fish and other animals, in which animal waste, in turn, is used to help support plant life. In particular, Biosphere II is a closed artificial ecosystem, in which plants, animals, and microbes, are used not only to provide food, but also to filter the air and water for reuse.

D. **Most agricultural waste can be recycled.** Composting is a process used to recycle waste. Bio-fuels, such as ethanol or methane, can be made from recycled wastes.

E. **Many processes used in agriculture require different procedures, products, or systems.** The procedures and products needed to plant a crop are different from those used in harvesting. For example, propagating and growing require plows and shovels, tractors with planting drills, and irrigation systems. In contrast, harvesting requires shears and hoes, hay thrashers, and bailing systems.

GRADES
6-8

In middle school, students should learn about and understand how technological inventions and innovations have helped to reduce labor hours and decrease the amount of land needed to grow crops and raise animals. For example, in 1930 it took five acres of land and 15-20 hours of labor to produce 100 bushels of wheat. In 1987, owing to the availability of better seed, improved fertilizer and pest control, and advances in farm machinery — the tractor, plow, planter, combine, and trucks, for example — 100 bushels of wheat could be produced on three acres of land with only three hours of labor.

At this level, students should broaden their understanding of the field of biotechnology — the manipulation of living things to make products that benefit human beings. In addition to combating disease and promoting human health, biotechnology is also being put to work in developing plants that resist diseases, increase crop yields, and reduce the use of pesticides. Using their research skills, students should investigate various biotechnology practices and assess the positive and negative consequences.

Students should continue to investigate the agricultural processes and systems used when planting, treating, harvesting, preparing, and utilizing the products for consumption. For example, students could design, develop, use, and assess a terrarium or hydroponics station that functions as part of a larger closed system supporting living organisms. They could manage the system and determine the rate, amount, and timing of light, water, nutrients, and waste recycling that the system requires. They should also be able to determine if the system is doing what it is supposed to in order to maintain life, scrub or clean the air and water, and assess the system for proper functioning. Students should troubleshoot and maintain the system if any part fails to function properly.

Opportunities to examine how agricultural waste is used and the trade-offs that are associated with waste recycling and other agricultural practices will help students understand how one action may cause an unexpected reaction. Improvements in fertilizers and pesticides have allowed large quantities of food to be grown. In some cases, however, these fertilizers, weed killers, and insecticides have mixed with groundwater and have contaminated the water supply. Students must be able to assess such pluses and minuses if they are to make reasonable decisions about the use of agriculture and related biotechnologies and generate solutions to remedy problems.

In addition, students should recognize that many systems are used in agriculture, including irrigation and agroforestry. Often these systems are designed to work in coordination with each other. For example, an agroforestry system, which is the coordinated use of plants, crops, and water resources, is often combined with an irrigation system to increase crop yield. Planting trees, shrubs, and crops in alternating plots, in addition to an irrigation system, will increase the diversity of soil usage, conserve energy, protect soil and water resources, and increase production.

GRADES
6-8

In order to select, use, and understand agricultural and related biotechnologies, students in grades 6-8 should learn that

F. **Technological advances in agriculture directly affect the time and number of people required to produce food for a large population.** New tools and machinery, such as milking machines, trucks, and combines are designed to make work easier and more productive. Today, fewer people are involved in producing food, while more people are needed for processing, packaging, and distributing it.

G. **A wide range of specialized equipment and practices is used to improve the production of food, fiber, fuel, and other useful products and in the care of animals.** For example, farmers use lasers to level their fields and the global positioning system (GPS) for precision farming. Wildlife habitats create special environments that encourage beneficial insects that in turn increase plant pollination and pest control.

H. **Biotechnology applies the principles of biology to create commercial products or processes.** Advances in the area of molecular genetic biotechnology have been made in the pharmaceutical industry (improved therapeutic drugs), agricultural industry (herbicide-resistant, pesticide resistant, and climate-adapted crops), as well as in medicine (gene therapies).

I. **Artificial ecosystems are human-made complexes that replicate some aspects of the natural environment.** For example, a terrarium is used to raise plants or animals in an enclosed habitat. The terrarium acts as a total environment using all the systems of life, such as food, water, shelter, and space. Managing an artificial ecosystem requires gathering data to plan, organize, and control processes, products, and systems. For example, operating a hydroponics system within a closed (or open) environment requires total control and cultivation. Temperature, nutrients, light, air circulation, and monitoring of insects all need management in order for the system to function properly.

J. **The development of refrigeration, freezing, dehydration, preservation, and irradiation provide long-term storage of food and reduce the health risks caused by tainted food.** For example, the process of irradiation involves bombarding food with low doses of high-frequency energy from gamma rays, X-rays, or accelerated electrons. The purpose of the process is to extend shelf life for weeks instead of days by inhibiting maturation and decay.

GRADES

9-12

In grades 9-12, students' understanding of the technologies used in agriculture can increasingly draw upon their knowledge of the underlying principles and concepts of technology, such as design and systems. Researching and working with different types of agricultural products and systems will help students make connections to other topics studied in technology and understand the value of the use of technology in agriculture.

Students may study the effects of waste and pollutants deposited in watersheds. They may also study the various methods of restoring a polluted watershed including bioremediation, which is the use of microbial organisms to degrade and detoxify pollutants. Students may test soil run-off for various pollutants and design and develop a system that might serve as a model for improving environmental conditions.

Students should discuss the need for regulations governing technologies used in agriculture. They also should discuss the social side effects and trade-offs of using various technologies in order to produce an abundant supply of better-tasting and more nutritious food. To reinforce their understanding, students could conduct research and present their findings on the positive and negative effects of a particular process, product, or system developed for agriculture. They could study the effects of genetically altered plants or of plants newly introduced to an area.

In order to select, use, and understand agricultural and related biotechnologies, students in grades 9-12 should learn that

K. **Agriculture includes a combination of businesses that use a wide array of products and systems to produce, process, and distribute food, fiber, fuel, chemical, and other useful products.** Crops (e.g., cotton, wheat, tobacco, and grains) and livestock (e.g., cattle, sheep, and poultry) are bought and sold by individuals, corporations, and financial institutions. Local, state, and federal governments regulate the marketing and safety of agriculture products and systems.

L. **Biotechnology has applications in such areas as agriculture, pharmaceuticals, food and beverages, medicine, energy, the environment, and genetic engineering.** Biological processes are used in combination with physical technologies to alter or modify materials, products, and organisms. Fermentation, bio-products, microbial applications, separation and purification techniques, and monitoring and growth processes are key examples of biotechnology applications. Selection of genetically modified seeds, application of modified organisms (i.e., ice-minus bacteria to prevent frost damage to plants), and uses of algal fertilizers generated from photobioreactors are good examples of extending agricultural practices through biotechnology applications.

M. **Conservation is the process of controlling soil erosion, reducing sediment in waterways, conserving water, and improving water quality.** For instance, terraces, used in gardens or on farmland, prevent erosion by shorten-

GRADES
9-12

ing the long slope of land into a series of shorter, more level steps. This allows heavy rains to soak into the soil rather than running off and causing erosion.

N. **The engineering design and management of agricultural systems require knowledge of artificial ecosystems and the effects of technological development on flora and fauna.** For example, wise water use for gardens or farmland involves considering plant needs and efficient watering methods before installing, using, and maintaining irrigation systems. Management of agriculture requires considering such topics as the amount, orientation, and distribution of crops and other plants, the effects of pests, and the management of land and animals to prevent fire or drought. For example, pest management involves managing agricultural infestations (including weeds, insects, and diseases) to reduce adverse effects on plant growth, crop production, and environmental resources.

VIGNETTE

Hydroponics System

This example provides students with an opportunity to focus on the application beyond growing foods to incorporating the use of plants and animals to complete a total environment. [This example highlights some elements of Grades 9-12 *Technology Content Standards* 2, 8, 9, 10, 11, and 15.]

Ms. M challenged her high school students to design a hydroponics system using plants to scrub the air and remove excess carbon dioxide. The system would need to provide enough oxygen in an underwater research station for a four-person crew. Working in groups, the students selected a variety of vegetables that thrive in a hydroponics environment and have excellent gas exchange properties. They determined the amount of space needed for each type of vegetable. One group conducted experiments to learn how to control the flow of water through tubes (PVC pipe) and regulate the distribution of nutrients to the plants. Another group tested pumps to lift water into reservoirs. The students invented devices to control the level of liquid in the reservoirs, and they experimented with lighting that would cause plants to grow and produce more efficiently. Several groups designed an apparatus to support the growing tubes, reservoirs, pumps, control devices, and lights. This design challenge facilitated student learning about plants, about how to increase productivity through the use of biotechnology applications, and provided an opportunity for them to explore the many uses of combined biological and physical technologies.

Once the groups had designed various hydroponics systems, they made and tested prototypes. Through this prototyping process, they discovered whether the plants would grow, if the space requirements were adequate, and if the fluid and electrical systems functioned properly. They were even able to sample the produce.

As an extension of this activity, Ms. M decided to introduce information-technology tools that could be used to convey the students' experiences to other students, school staff, school board, parents, and the community. Some students chose to use the World Wide Web; some designed and made video programs; and others wrote illustrated articles in the school newspaper.

STANDARD

Students will develop an understanding of and be able to select and use energy and power technologies.

Energy is the ability to do work. Large supplies of energy are a fundamental requirement of the technological world. Although energy and power are often used interchangeably, they should not be — each has unique characteristics that differentiate one from the other. Energy is the capacity to do work. Power may be defined as the rate at which energy is transformed from one form to another or as the rate at which work is done.

Technological products and systems need energy that is plentiful, cheap, and easy to control. Thus, the processing and controlling of energy resources, often called fuels, have been key features in the development of technology.

Energy drives the technological products and systems used by society. The quality of life is sometimes associated with the amount of energy used by society. Choices about which form of energy to use influence our society and the environment in various ways. There are always trade-offs to consider in the source of energy that may be chosen. Energy and power systems can pollute the environment. Some sources of energy are non-renewable — once they are used, they will no longer be available — while others are renewable, such as fuel made from corn. Many of our energy needs are met by burning fossil fuels. Nuclear energy provides a source with less air pollution and no carbon-dioxide buildup, but nuclear waste is more dangerous for longer periods of time than waste from other energy sources.

It is the responsibility of all citizens to conserve energy resources to ensure that future generations will have access to these natural resources. To decide what energy resources should be further developed, people must critically evaluate the positive and negative impacts of the use of various energy resources on the environment.

GRADES

K-2

In grades K-2, students will investigate the particular types of energy that they are most likely to encounter. Energy enables plants to grow, cars to run, and food to be cooked. Energy comes from many different sources in nature, such as fossil fuels and the sun. Many toys and household items would not work without electricity, fuels, or other forms of energy. Students at this early age need to be exposed to many different sources of energy to begin to develop an understanding of this complicated topic. They also should begin to understand the relationships between energy, power, and work. Safety should be stressed with young children in working with energy and power.

Conservation is a very important concept when working with energy resources, and students should learn to avoid wasting energy in their schools and homes. For example, students should learn to turn lights off when they are leaving a room and to turn the television off when they are no longer watching it.

In order to select, use, and understand energy and power technologies, students in grades K-2 should learn that

A. **Energy comes in many forms.** It is used to do work. An early source of energy for machines was provided by human or animal muscle and was converted from food that was eaten. A car engine changes chemical energy (gasoline) to mechanical energy (motion). Many appliances in the home and school use electrical energy.

B. **Energy should not be wasted.** Toys and appliances should be turned off when they are not being used. Many energy resources, often called fuels, that are used to heat and light our homes, run our cars, and cook our food are nonrenewable. There is a limited supply of these resources, and the supply is being used up.

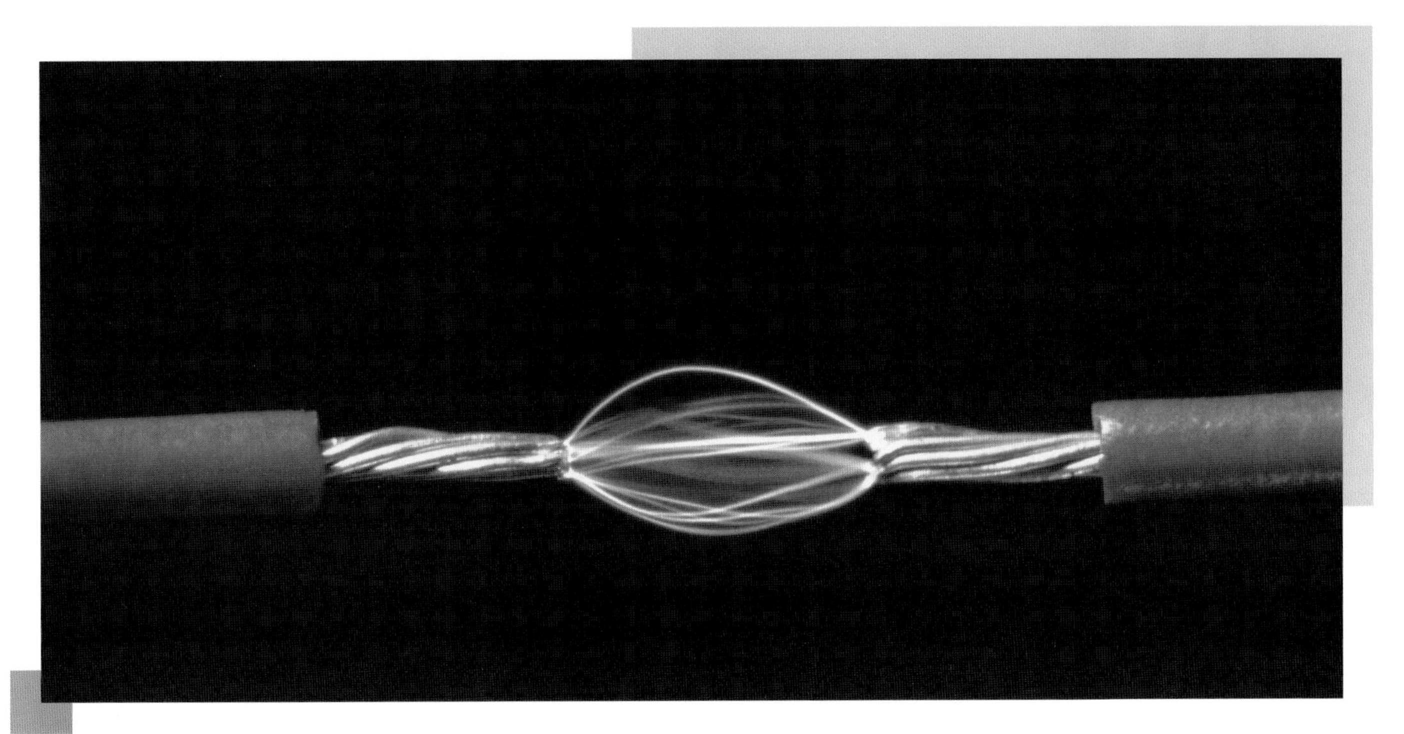

GRADES
3-5

Energy is behind every movement and change in the world. Energy comes in many different forms, such as thermal, radiant (light), electrical, chemical, mechanical, and others. Many technological devices are driven by energy. Power systems, such as gasoline engines, generators, and solar cells, transform different forms of energy in order to do work. In grades 3-5, students will continue to study energy and how these ideas are related.

Students should also learn how tools, machines, products, and systems need energy sources for their operation. Power systems are used to convert energy to work that results in some thermal losses. Examples of these conversion machines include electric generators (mechanical to electrical), electric motors (electrical to mechanical), dry cell batteries (chemical to electrical), home furnaces (chemical to thermal), automobile engines (chemical to thermal to mechanical), incandescent lamps (electrical to radiant), and solar cells (radiant to electrical). Students should also learn the proper procedure for plugging an appliance into an electrical outlet, and they should be taught to be careful of the socket when screwing in a light bulb.

Energy should be used carefully so that it is not wasted. In elementary school, students should learn how to conserve energy, which is an important concept in technological literacy. Students can take an active role in conserving energy by riding their bikes to school instead of riding in a car and by using a fan instead of air conditioning. They can also investigate how individuals who make products and systems can conserve energy.

In order to select, use, and understand energy and power technologies, students in grades 3-5 should learn that

C. **Energy comes in different forms.** Forms of energy include thermal, radiant (light), chemical, mechanical, electrical, and others. Some energy sources cost less than others, and some give off less pollution. Electrical energy is used in an electric motor, and solar cells can be used to transform solar energy to electrical energy to operate a calculator.

D. **Tools, machines, products, and systems use energy in order to do work.** A well-designed tool, machine, product, or system minimizes energy losses. For example, machines should be designed to apply energy efficiently to do a useful task. Energy is an important resource in technology.

VIGNETTE Developing and Producing a Product or System Collaboratively

This vignette presents some activities related to energy and power. The students study technology-related problems that can be accomplished in elementary school classes. [This vignette highlights some elements of the Grades 3-5 *Technology Content Standards* that provide connections with Standards 3, 8, 9, 11, and 16]

While studying about energy and power systems and reading books that discuss robots, haunted houses, and how things work, students were challenged to solve one of the following problems: (1) use age-appropriate circuit diagrams with proper assistance to make a robot out of reclaimed materials that includes at least one series circuit and provides light; (2) make a construction paper model of a haunted house that includes at least one series and parallel circuit and lights up two different rooms; or (3) build a flashlight that works, using the following components: size 'C' battery, wire, light socket, light bulb, paper towel tube, construction paper, wire clips, and tape.

After designing, developing, and producing their solutions, the students should discuss their solutions and evaluate their products and systems to see how closely their results fit the requirements.

GRADES
6-8

At the middle school level, students will learn about the concepts of energy, power, and work. They will engage in hands-on experiences in converting energy from one form to another and in learning that energy can be transmitted from one location to another.

There are many processes that are used to transform energy into useful work. Fossil fuels, such as oil, can be burned to produce thermal energy, which in turn is used to boil water and produce steam that, in turn, is fed into a steam engine to propel a locomotive. Water falling over a dam can be converted into mechanical energy, which turns a generator that produces electricity for lighting and other conveniences in a home. To understand these processes, students should design and build different devices that use energy to drive a product or system. They should then test their devices in order to determine how efficiently they are working.

If not used wisely, nonrenewable energy sources can be depleted too rapidly. Because of these concerns, conservation and the search for alternative energy sources have become important priorities for society. Students should investigate and discuss various methods for conserving energy. They can then build models of their ideas and test them. Students also could design, build, and test alternative energy devices.

In their study of energy students will come in contact with many devices and tools that require special care in the way they are used, handled, and stored. The development of safe work habits cannot be overemphasized. It is essential that students learn the proper safety procedures when working with these energy technologies.

In order to select, use, and understand energy and power technologies, students in grades 6-8 should learn that

E. **Energy is the capacity to do work.** Energy is required for a broad range of actions, from walking to running a diesel engine. Energy is an important input to many technological systems. Work is the product of force multiplied by the distance through which the force acted. Work is measured in foot-pounds in the English system and in Newton-meters, or joules, in the metric system.

F. **Energy can be used to do work, using many processes.** For example, electricity can be generated by using geothermal energy to turn a turbine, which subsequently turns a generator to produce an electrical voltage. Another example involves an internal combustion engine: gasoline vapor is combined with air and ignited with a spark plug; the spark plug explodes inside the cylinder creating high pressure and temperature; the pressure acting on the piston pushes it down; the piston is connected to a piston rod that turns the crankshaft.

G. **Power is the rate at which energy is converted from one form to another or transferred from one place to another, or the rate at which work is done.** Power is calculated by dividing the energy provided by the time taken to provide it. Common power measurements are horsepower and kilowatt. An example of the difference between the concept of energy (or work) and power can be seen in a student climbing a set of

7

stairs. To climb from one floor of a building to another takes the same amount of energy to do the same work no matter how fast the student climbs. However, to climb twenty stairs in 30 seconds is quite different from climbing the same twenty stairs in 10 seconds. Climbing faster requires the same amount of energy but *more* power, in the previous example three times more power.

H. **Power systems are used to drive and provide propulsion to other technological products and systems.** A portable generator, for example, can be used to provide electricity to remote dwellings.

I. **Much of the energy used in our environment is not used efficiently.** Conservation is the act of making better use of energy. Individuals can conserve energy by car pooling, driving the speed limit, and turning off lights. Builders can conserve energy by installing better insulation, and manufacturers can conserve energy by building more energy-efficient products. The rate at which energy is being used in the world is increasing. This rapid increase has created a concern that natural resources may be depleted in the future before other energy resources are available to replace them.

GRADES

9-12

Students in grades 9-12 will study energy, power, and work concurrently in their science and technology classes. They should synthesize the concepts and principles learned in science with the knowledge gained in the study of technology to achieve a well-rounded understanding of energy and power.

Energy, which is the capacity to do work, can be converted from one form to another. Thermal energy is usually a by-product in a conversion process. Some energy converters are more efficient than others. For example, electric generators are more than 95 percent efficient at converting mechanical energy to electrical energy, while the fluorescent lamp is only about 20 percent efficient in converting electrical energy to radiant energy. However, the fluorescent lamp is more than four times as efficient as the incandescent lamp.

Energy can be classified into two types — kinetic (energy associated with motion) and potential (stored energy). Energy can come from a number of resources in nature, such as the sun (radiant), from tides or falling water (mechanical), from the burning of fuels (chemical to thermal), and from chemicals such as those used in batteries (chemical to electrical). Students can work with devices that convert one form of energy to another such as electrical to mechanical (motor), electrical to radiant (lamp), or mechanical to electrical (wind generator).

The United States is among the most highly developed countries in the world, and Americans use an ever-increasing amount of energy, even though much of it is derived from nonrenewable resources. Because many environmental and societal concerns are associated with the proper use of energy, research and development are underway to test alternative and renewable resources. Students should investigate our dependence on fossil fuels, the use of alternative sources of energy, and the trade-offs associated with each.

All power systems have inputs, processes, outputs, and typically some type of feedback. Students should research energy inputs (e.g., thermal, chemical, nuclear, mechanical, radiant, and electrical), processes (e.g., conversion, transmission, and storage), and outputs (work and thermal loss). At this level, students can learn about the various sources of energy, the influence of energy and power on society, and energy and power systems. Students should be exposed to the Second Law of Thermodynamics.

Conservation is also important in managing energy sources. Conservation can take various forms, from simply turning off appliances when they are not being used to designing products that are more energy efficient. Students should investigate various approaches to conserving energy. For example, they could investigate recycling materials instead of producing new ones. When new energy and power systems are designed, conservation of energy and environmental concerns must be incorporated. Students should investigate the by-products of systems, such as the waste stream in the nuclear fuel cycle. Using this information, they could then design, develop, and test power systems and determine if they are efficient and non-polluting.

In order to select, use, and understand energy and power technologies, students in grades 9-12 should learn that

J. **Energy cannot be created nor destroyed; however, it can be converted from one form to another.** In scientific terms, this is called the Law of Conservation of Energy, which can be stated as "The total energy of an isolated system does not change." Understanding scientific concepts and laws concerning energy is necessary in order to develop technologies for utilizing energy. These concepts and laws describe the nature of energy. Energy can be classified as either kinetic or potential. Kinetic energy is the energy a body has associated with its motion. Potential energy is energy a body has because of its position (if it can be acted upon by a force) or condition; it is often referred to as stored energy.

K. **Energy can be grouped into major forms: thermal, radiant, electrical, mechanical, chemical, nuclear, and others.** Some forms of energy cannot be transported easily. In transporting or transmitting energy, losses from the source of energy to the destination occur. Many times technology systems that use a great deal of energy are located near the energy source. An example of this is an electric generating plant located near a source of energy, such as a coal mine. The combustion of fossil fuels (e.g., coal, natural gas, and petroleum) provides one of the largest sources of energy today.

L. **It is impossible to build an engine to perform work that does not exhaust thermal energy to the surroundings.** This is one form of the "Second Law of Thermodynamics." No energy system can be 100% efficient. Large coal-fired electric generation systems strive for 40% efficiencies. That means that 60% of the energy from the coal is lost in the form of heating the environment rather than being turned into electrical energy. The law also has many wide ranging consequences, such as the fact that there can be no perpetual motion machine.

M. **Energy resources can be renewable or nonrenewable.** Examples of renewable resources are the sun and agricultural products, while nonrenewable resources include fossil fuels, such as coal, oil, and natural gas. Alternate and sustainable energy resources are being developed and tested in order to replace or supplement nonrenewable sources. For example, garbage can be used to produce methane gas and then burned for thermal energy. Also, corn can be fermented to produce ethanol (grain alcohol), which then can be used as a fuel. Power systems should be designed to conserve energy and to provide maximum efficiency with minimal environmental degradation. For example, aircraft manufacturers are making more energy-efficient engines. Waste products associated with power systems can pollute the natural environment.

N. **Power systems must have a source of energy, a process, and loads.** Usually feedback is part of this system. For example, the output of the system is sampled and provides a signal back to the input or process phase of the system in order to modify it. Power systems convert energy from one form to another and may transfer energy from one place to another. An example would be to burn coal in order to heat water and make steam, which turns a turbine and ultimately generates electricity.

STANDARD

17 Students will develop an understanding of and be able to select and use information and communication technologies.

People have long used various technologies to communicate over distances. The invention of movable type provided the means for a transfer of knowledge to people all over the world. Although writing and printing have become visual means for communication, people did not typically consider them to be communications technologies, viewing them simply as technologies that met a particular need. The book was viewed as not having much in common with the telephone, or the phonograph with the fax machine. In the past couple of decades, such thinking has changed dramatically. Technologies that record, store, manipulate, analyze, and transmit data have developed into important areas of study worthy of being considered equally with other technologies.

The change has come with the recording and storage of all sorts of data in the same digital form as "bits" — strings of zeros and ones, or offs and ons — that can represent letters and numbers, colors on a computer monitor, notes in a Beethoven sonata, and many other types of information. Modern telephone companies, for example, transform the sounds of a telephone conversation into bits that are sent through fiber-optic cables in exactly the same way as data is sent from computer to computer. Data, information, and knowledge have become the fuel that runs the communication technology engine. Information and communication technologies include computers and related devices, graphic media, electronic transmitters and receiving devices, entertainment products, and various other systems.

Powerful technologies that deal with information in a digital form — computers, data-storage devices, fiber-optic communications, and others — have revolutionized society's information-handling capacity and led to the current era being called the Information Age. Information itself is a valuable commodity, which has become more widely available than in the past.

GRADES
K-2

By the time children enter kindergarten, they will already have had many informal experiences in using technology to communicate with their friends and family, as well as to find answers to their questions. For example, many students will have looked at picture books and used the Internet for entertainment and information. Additionally, they may have experienced various entertainment activities, such as going to the movies, watching television, or using a telephone or a DVD player.

The K-2 laboratory-classroom will provide formal opportunities for students to learn about the communication process and the different ways that they can locate information and communicate with others. At this level, students will learn that information is data that has been organized. In later grades, students will build on this concept and expand their vocabularies.

Information and communication tools and systems are available in many different forms. Computers are at the center of the Information Age and are the primary tools used. Students should be able to operate computers to perform simple tasks, such as writing, learning basic operations, communicating with others, and making graphic images. In addition, they should be able to use other information technology tools to locate information from both print and electronic sources. These experiences will reinforce their understanding of how communication systems work, how they can be used as an entertainment medium, and how they make communicating with others and gathering information easier.

Information and communication technologies use a specialized vocabulary that is important to understand, including words, symbols, and pictures. As students learn the alphabet and numbers, they need to realize their importance as symbols. Language is key to communication.

In order to select, use, and understand information and communication technologies, students in grades K-2 should learn that

A. **Information is data that has been organized.** Data includes such things as numbers, amounts, words, symbols, sounds, and images.

B. **Technology enables people to communicate by sending and receiving information over a distance.** Communication systems help people to communicate information better and to locate information more easily. The telephone is a good example of a technology that improves communication all over the world.

C. **People use symbols when they communicate by technology.** These symbols are a part of the language of technology. For example, a stop sign is a type of symbol. Small pictures or icons on a computer screen are other types of symbols.

GRADES

3-5

In grades 3-5, the study of information and communication technologies can be invaluable in helping students better understand what they are learning in language arts, science, mathematics, and social studies. It also will help students in developing their research and communication skills, which are valuable components of every subject area.

The use of information and communication systems can enhance human knowledge and productivity, as well as provide entertainment. In grades 3-5, students will learn how facts, data, information, and knowledge are interrelated. To reinforce these ideas, students will have hands-on experiences in accessing facts, processing data, organizing these facts and data into information, and then interpreting the information to produce knowledge. The computer provides an important way to access facts and data and then to convert them into information. For example, the students could use the Internet to collect information about how technology has been used to improve agricultural production. Once they have collected their information, they could use a simple spreadsheet program to show how agricultural production has increased over the years. This information can then be used for writing a research paper.

Communication is the exchange of information among people over a distance. Communication technology is the transfer of information between people and/or machines through the use of technology. Students at this grade level should be given various opportunities to use information and communication tools in order to experience firsthand how technology can be used to enhance the communication process, to help access information, and to provide entertainment activities. For example, in a unit of study about the solar system, students could use a computer to create a graphic representation of the planets, or they could apply their building skills to make a model of the stars. They then can use these representations when they present what they have learned to their classmates and teachers. Other types of communication tools that students may use include digital or still cameras, video cameras, audio recording devices (e.g., tape recorders), World Wide Web pages, and spreadsheets.

Symbols, letters, numbers, and icons are used to represent data. For instance, a popular symbol is a √ (check), which means, "correct" or "okay." The letters of an alphabet are combined to create words. Icons are small pictures that represent information, such as a computer function. Numbers are used to represent a quantity, to identify a location, to identify a specific object in a collection, to name something, or to represent a measurement.

In order to be able to select, use, and understand information and communication technologies, students in grades 3-5 should learn that

D. The processing of information through the use of technology can be used to help humans make decisions and solve problems. Computers can be used to record and store data, to access data easily, and to provide a means to display and manipulate it.

E. Information can be acquired and sent through a variety of technological sources, including print and electronic media. Computers can be used very efficiently to store, retrieve, and process information. A growing number of people work in jobs related to the processing and distributing of information.

F. Communication technology is the transfer of messages among people and/or machines over distances through the use of technology. Communication systems, such as the telephone, electronic mail, and television, are used to improve the communication process.

G. Letters, characters, icons, and signs are symbols that represent ideas, quantities, elements, and operations. For example, "+" and "-" are used to indicate addition and subtraction; an up arrow on a map refers to the north; and a red octagonal sign means "Stop." Symbols, measurements, and sketches represent information.

GRADES

6-8

Information and communication technologies have become an important part of everyday life. However, what happens behind the scenes when a telephone call is made or an e-mail message is sent is often puzzling. People simply take for granted that the message will arrive at the correct destination. Students should develop an understanding of how information and communication systems function.

Information and communication systems allow information to be transferred from humans to humans, humans to machines, and machines to humans. These systems enable people to gather and process data and information more easily and, therefore, to communicate more effectively.

Information and communication systems assist humans in making decisions and solving problems. Students should not interpret all communication messages as being true. They should research and check out messages to determine the facts.

Entertainment has also been enhanced by various information and communication technologies. Radio, television, movies, and video games are all products of technology. Students should explore the historical development of various forms of entertainment and then project how they think entertainment will change or stay the same.

At the middle school level, students should explore the different steps in the communication process. First, a message must be encoded and then transmitted or switched through a channel (wire, fiber optics, etc.), and finally received and decoded by the receiver. In order to understand this process, students should design and send messages using various types of communication systems, paying close attention to each step of the process.

The intended audience, the medium that is used, the purpose of the message, and the nature of the message influence the design of a communication. As a result, information must be closely evaluated according to its source, content, purpose, and intent in order to determine its value. Students should explore the different factors that influence a message. They then can apply this information to help them define a set of requirements in order to assess the information that they send or receive through information systems.

In grades 6-8, students should be provided with numerous opportunities to use information and communication systems for assistance in solving problems, making better decisions, and communicating with others. Students could use information systems to gather facts on technologies that have positively affected society, the effect of medical technologies on increasing life spans around the globe, or how information technologies have made it difficult for totalitarian states to maintain a stronghold on their citizens. Students should then organize and maintain their information in a systematic manner. After analyzing the information that they have collected, students can communicate their findings to their classmates.

The use of symbols in technology has become commonplace in today's society. These symbols represent measurements, terms, and ideas. Drawings are graphic representations of objects in either two or three dimensions. Students should gain experiences in the language of technology to express themselves and to communicate to others.

In order to select, use, and understand information and communication technologies, students in grades 6-8 should learn that

H. Information and communication systems allow information to be transferred from human to human, human to machine, and machine to human. People create information and communication technology systems to gather data, manipulate, and communicate information more effectively. Information is transmitted and received using various systems (e.g., telecommunications, digital, and printed). Transmission involves sending signals in a form, such as electro-magnetic waves or fiber-optic cable, that can travel over a distance.

I. Communication systems are made up of a source, encoder, transmitter, receiver, decoder, and destination. A communication system is similar to other systems in that it includes input, processes, outputs, and sometimes feedback. Information is encoded using symbols and graphics. "To encode" means to change the form of a message (as in pushing a key on a keyboard to produce a binary signal or changing a signal from analog to digital). Information must be decoded in order to be understood by the receiver. "Decoding" is the reverse of encoding, with data being converted back to symbols and graphics. Switching circuits allow signals to be sent back and forth in the communication process. A network is a system connected by communication lines to move information from one device to another. An example of a network is a local area network (LAN), which connects computers to a server. Computers are the primary tools used for networking information and communication technologies.

J. The design of a message is influenced by such factors as the intended audience, medium, purpose, and nature of the message. These factors should be taken into account when the message is created and transmitted to a particular audience. Communication-technology systems enhance the ability of handicapped people to communicate. Some examples include audiotapes, the Internet, and closed-captioned television.

K. The use of symbols, measurements, and drawings promotes clear communication by providing a common language to express ideas. Technological systems use specialized symbols and terminology. For example, an engineer uses specific symbols to represent doorway openings, pipe openings, and road widths. Symbols or icons are used on many computers, elevators, and telephones — the pound sign, the asterisk, and the letter "x," for example — to represent ideas and to communicate what should be done when the symbol or icon is pressed or used.

VIGNETTE

This example presents a problem-solving activity to develop a means to communicate to the public information about a local industry. [This vignette highlights some elements of the Grades 6-8 *Technology Content Standards* that provides connections with Standards 1, 3, 4, 6, 8, 9, 11, 17, and 19.]

Communicating Through a Home Page on the Web

Very few students at Northwood Middle School had any knowledge of one of the fastest growing activities in their area — the production of home pages for the World Wide Web. For this reason, Mr. M wanted to give his students experience in designing and putting a home page on-line.

The students began their study by contacting local Internet service providers to find out about what they did. After the research phase, Mr. M gave the students a design brief that outlined the problem statement, objectives, software and hardware, requirements, time frame, and evaluation method. They were challenged to design and place into operation a home page for a local children's clothing store. The new home page should convey the products of the company in an exciting, creative, and appealing way.

The students brainstormed various ideas and then produced sketches for a layout of the home page. Each sketch included graphics, a slogan, a color scheme, and special products. The students selected the design they liked best, and Mr. M reviewed the home pages and offered suggestions for improvements.

Mr. M, along with input from the store owner, then evaluated the students' work — how well the home page communicated the message, the quality and variety of planning used by the students, the overall creativity of the design, the quality of the work, in addition to whether the students met the requirements of the design and completed the task on time. As a result of this activity, students were able to design, develop, and use a communication technology.

GRADES
9-12

By grades 9-12, students should be comfortable with terms such as facts, data, information, and knowledge, and they should understand the relationships among them. Additionally, students should understand the processing and management of information, which will assist them in sending and receiving information.

The classic communication system is made up of an information source, an encoder, a transmitter, a receiver, a decoder, and an information destination. Feedback may be included in this process as well. Because noise is any unwanted signal that can interrupt or interfere with the communication process, students should investigate various methods to overcome noise and promote clear communication.

Messages are influenced by many factors, such as timing, sequencing, and processing. Because people today are bombarded daily with numerous messages, the usefulness of information depends on such factors as relevancy, timeliness, truth, completeness, and cultural value. The knowledge and information that is provided through information and communication systems can help to inform people, shape their personal views and concepts of reality, or to entertain them. At this level, students should experience activities in designing, using, and assessing with many different types of information and communication systems, including television, telephones, the Internet, data-processing systems, fiber-optic cable systems, and graphic-communication systems. They should know the purpose of each system and be able to select the best one for a given situation.

Because information has become such a valued commodity in today's society, many commercial companies are involved in information and communication technologies. Students can research, synthesize, and transmit messages to the public using mass media. They can also evaluate the quality of information that is received by using such techniques as comparing and contrasting information sources and examining the relevancy of the message.

Graphic-communication systems involve visual messages, such as words and pictures — newspapers, magazines, and print media exemplify this type of communication. Entertainment, including television, movies, videotape, music, and compact discs, is also a growing area of communication.

Symbols, measurement, conventions, icons, and graphic images are recognized components in the language of technology that are used to communicate messages. Students should communicate to others using the language of technology.

In order to select, use, and understand information and communication technologies, students in grades 9-12 should learn that

- L. **Information and communication technologies include the inputs, processes, and outputs associated with sending and receiving information.** All of these parts are necessary if information is to be shared and understood by the sender and receiver.
- M. **Information and communication systems allow information to be transferred from human to human,**

human to machine, machine to human, and machine to machine. Examples of these are: a) two people talking to each other over the telephone, b) a person inputting data in a computer using a keyboard, c) an electric fax machine providing a copy of a message to a person, and d) an automated system transferring financial records from one bank computer to another bank computer.

N. **Information and communication systems can be used to inform, persuade, entertain, control, manage, and educate.** Examples of such systems include the Internet, telephones, televisions, radios, computers, and fax machines. Information and communication systems are widely used in commercial endeavors to assist in decision-making and problem solving. Entertainment, which has been enhanced through technology, provides pleasure and enjoyment for people in their free time. The overall usefulness of information is dependent upon such factors as its relevance, timeliness, truth, completeness, and cultural value. These factors help shape the meaning of the information, which has become a valued commodity in today's society.

O. **Communication systems are made up of source, encoder, transmitter, receiver, decoder, storage, retrieval, and destination.** Noise, an unwanted signal that can interfere with the communication process, can interrupt the signal at any point in the process. Data and information can be stored to be retrieved later. Storage devices include CD-ROMs, hard drives, flash memory, and memory chips.

P. **There are many ways to communicate information, such as graphic and electronic means.** Graphic-communication systems involve the design, development, and production of visual messages. Examples of graphic systems include printing and photochemical processes, while examples of electronic systems are computers, DVD players, digital audiotapes, and telephones. This information can be expressed in various forms: electrical information can be formatted as digital (discrete bits) or analog (continuously variable signals). Multimedia combines information from a number of formats (audio, video, and data) and then transmits it. Television studios and telephone companies exemplify businesses that deal with multimedia.

Q. **Technological knowledge and processes are communicated using symbols, measurement, conventions, icons, graphic images, and languages that incorporate a variety of visual, auditory, and tactile stimuli.** For example, the international symbols developed for transportation systems have helped to communicate critical information to travelers: a circle with a slash represents "No" or "Do not do." Emerging technologies often generate new symbols, measurement systems, and terminology. For example, ;-) is a symbol used in e-mail and on-line chat rooms to represent a wink. The development of the computer has spurred new terminology, such as gigabyte (a unit of computer storage capacity equal to one billion bytes) and nanosecond (one billionth of a second).

STANDARD

18 Students will develop an understanding of and be able to select and use transportation technologies.

People view transportation as one of life's basic needs. Transportation systems take individuals to work, offer them convenient access to shopping, allow them to visit with their friends and family, provide opportunities for recreation, and carry all the material goods of a technological society.

The transportation system is a complex network of interconnected components that operate on land, on water, in the air, and in space. Although traveling into space has been realized, it has not yet become a fully integrated part of the larger transportation system. Many of the subsystems of the transportation system, such as highways, ports, airports, and others, are dependent upon other subsystems, and each in turn is made up of yet smaller components that are themselves interlinked and interdependent. Various forms of transportation have been in use for many years by a wide assortment of people, such as ships, boats, jets, helicopters, elevators, and escalators, while newer forms of transportation are used in limited areas or are still in experimental stages — magnetic-levitation trains and smart highways, for example.

The more complex life and work become, the more indispensable the elements of transportation systems are. Throughout history, transportation systems have brought different parts of the world closer together. In the early twentieth century, for example, a plane flight across the United States would take approximately 26 hours. Now, through advances in technology and improvements in the aviation system, the same trip can be accomplished in six hours or less. If the Concorde were allowed to fly within the continental United States, it would take just two hours to travel from one side of the country to the other, and the space shuttle, once in orbit, makes the same trip in only nine minutes.

Because transportation has become such an integral part of life, people often take it for granted or consider it an ordinary part of the world. As transportation has advanced, society has become increasingly dependent upon cars, highways, and other aspects of travel. Too often little heed is paid to the environmental consequences or to the effects of rapid expansion that has accompanied transportation improvements. Future use of transportation systems should take into account ways to reduce energy consumption and air pollution, while promoting economic development and supporting international commerce.

GRADES

K-2

Children at this level have experienced various forms of transportation in their lives, and they typically consider transportation only in terms of the individual devices, such as cars, buses, trains, or planes. They know that a car uses roads and highways, but they do not think of the roads and highways as part of a larger road system, nor do they understand how the highway system works within the entire transportation system. Students need to learn how transportation systems work and how to use those systems safely and appropriately. Through experiences in grades K-2, students will begin to learn how various parts of the transportation system work together, a concept that will be further expanded in later grades.

Students should understand that caring for transportation vehicles is an important part of the process of how a transportation system and its various parts work. For example, a class could explore how animals travel from place to place and then relate that movement to how students move from home to school and back. The students could discuss how a vehicle could be cared for and what parts could break down. In addition, students should design, make, and use models of various vehicles, such as cars, boats, and planes, and discuss how the vehicles are used in different environments to move individuals and goods.

In order to select, use, and understand transportation technologies, students in grades K-2 should learn that

A. **A transportation system has many parts that work together to help people travel.** The roadway, vehicles, fuel, and controlling signs are just a few of the parts in a transportation system. Understanding how a transportation system works helps people use it properly, such as walking on the left side of the street facing traffic when sidewalks are unavailable.

B. **Vehicles move people or goods from one place to another in water, air, or space and on land.** People's needs and wants influence the design of a transportation device, vehicle, fuel, and system. For example, cars replaced the horse and buggy because they allowed people to move faster. Goods are often moved in specially designed carriers, such as in refrigerated containers, on conveyor belts, or through piping systems.

C. **Transportation vehicles need to be cared for to prolong their use.** People sometimes keep a log of what they must do to care for a vehicle, such as keeping it clean, rotating the tires, and looking for damage.

GRADES

3-5

At this level, children can begin to see transportation as a system in which all the parts work together to help move people and goods from place to place. Examples of transportation systems that students may explore include railways, waterways, roadways, and airways. Components of these systems may involve elevators, conveyor belts, pipelines, cars, ships, planes, refineries, and gas stations.

Students will design, make, use, and assess a simple transportation system in order to understand how it works and how it is designed for a special purpose. An example of this is a people-mover system to be used in the school. In addition, they should study how a transportation system, such as railways or highways, is comprised of subsystems and how these subsystems act as the input, process, output, or feedback for a larger transportation system.

Because balloons appeal to many children, a hot air balloon activity could be used as an introduction to explore how air transportation has changed throughout history. The students could learn about the development of various air transportation vehicles and find out how a hot air balloon moves through the air. They could then design, make, and test a model of a hot air balloon. Using their knowledge of how things work in relation to what they have learned in science, mathematics, social studies, art, and language arts, students will recognize that the transportation system is a complex arrangement of many subsystems and that it requires large amounts of energy to operate.

In order to select, use, and understand transportation technologies, students in grades 3-5 should learn that

D. **The use of transportation allows people and goods to be moved from place to place.** The development of transportation systems has had a significant influence on where people live and work.

E. **A transportation system may lose efficiency or fail if one part is missing or malfunctioning or if a subsystem is not working.** For instance, an accident on a highway can throw a whole traffic pattern into chaos. Severe thunderstorms over Atlanta can result in the cancellation of airline flights up and down the east coast of North America.

GRADES

6-8

Students will explore and learn how various methods of transportation are used in the environments of land, water, air, and space. Each environment requires specialized vehicles and systems for moving people and goods. A skyscraper, for example, employs elevators and escalators to move people up and down within the building.

Asking what might happen if a particular subsystem were not working or missing could lead students to reflect on the interdependence of systems in transportation, as well as the relationships of those systems to other systems. Students will be able to recognize the different subsystems of the transportation system (e.g., structural, propulsion, suspension, guidance, control, and support) and recognize how they work together. To increase their understanding of these subsystems, students may design and develop models of them. For example, the structural subsystem includes the framework and body of a vehicle. Students should design and develop a model of a new vehicle to be used on land, in the sea, in the air or in space in order to see firsthand how the structural subsystem is related to the environment in which the subsystem is used.

Finally, to develop an appreciation of how transportation systems can be adjusted or modified to help the environment, students could study the environmental consequences of using various alternative fuel sources.

In order to select, use, and understand transportation technologies, students in grades 6-8 should learn that

F. **Transporting people and goods involves a combination of individuals and vehicles.** For example, the movement of a product from one part of the country to another may involve the person shipping the item, a delivery truck, a bus, plane, or train, the people involved in controlling the product's location, as well as those who made the road, the car, and the fuel.

G. **Transportation vehicles are made up of subsystems, such as structural, propulsion, suspension, guidance, control, and support, that must function together for a system to work effectively.** Structural systems are the framework and body of a transportation vehicle or system. Propulsion systems provide the energy source, energy converter, and power transmitter to move a vehicle. Suspension systems connect or associate a vehicle with its environment. Guidance systems provide information to the operator of a vehicle. Control systems receive information from the guidance system to determine the changes in speed, direction, or altitude of a vehicle. Support systems provide life, legal, operational, maintenance, and economic support for safe and efficient operation.

GRADES
6-8

H. Governmental regulations often influence the design and operation of transportation systems. State agencies regulate the use of highway systems, set speed limits, and control other operating conditions. The Federal Aviation Administration regulates airspace and air safety and issues licenses to pilots.

I. Processes, such as receiving, holding, storing, loading, moving, unloading, delivering, evaluating, marketing, managing, communicating, and using conventions are necessary for the entire transportation system to operate efficiently. These processes may be used individually or in various combinations to move goods and people. For example, a conveyor system uses many of these processes to move boxes of goods in stages from one location to another.

GRADES
9-12

In grades 9-12, students' understanding of the transportation system will expand to encompass the entire concept of intermodalism, which provides a seamless and an effective method to move people and goods. They will also learn about the vital role that transportation plays in manufacturing, construction, communication, health and safety, recreation and entertainment, and agriculture. For example, the movement of goods in just-in-time (JIT) manufacturing is directly dependent on the global transportation system. Many industries use materials and prefabricated parts from other countries or from other parts of the country. These goods arrive just as they are needed (instead of being stored and used at a later time) to be used to manufacture products, such as cars and clothing. The transportation system is key to the use of JIT manufacturing, which helps in the reduction of storage needs and resource costs.

Although concepts can be learned from reading and discussion, students should have direct experiences with designing, developing, using, and assessing various transportation systems to understand them thoroughly. For example, students could design and develop a bicycle pathway, which would offer cyclists a safer alternative to congested streets. Students also could examine and develop examples of intelligent transportation systems. These systems integrate such technologies as computers, electronics, communications gear, and safety devices in order to make traveling more efficient and safe.

Classroom discussions should include such transportation issues as pollution, congestion, accidents, and fuel consumption. These issues should inspire students to devise solutions or innovations to solve the problems. Students, for example, could design, develop, operate, and assess an improved transportation system for moving people, which takes into account such factors as speed, cost, safety, and environmental impacts.

In order to select, use, and understand transportation technologies, students in grades 9-12 should learn that

J. **Transportation plays a vital role in the operation of other technologies, such as manufacturing, construction, communication, health and safety, and agriculture.** The transportation system includes the subsystems of aviation, rail transportation, water transportation, pedestrian walkways, and roadways. Each subsystem uses a wide array of devices, vehicles, and systems in order to move people, materials, and goods.

K. **Intermodalism is the use of different modes of transportation, such as highways, railways, and waterways as part of an interconnected system that can move people and goods easily from one mode to another.** An example of intermodalism is a semi-truck container that is hauled on an ocean cargo ship from another country, transported to a railcar, and finally, attached to a semi-truck that travels a highway to deliver goods. The same process is used by people who travel to all parts of the world using different modes of travel, from airplanes to ships, to buses, trains, or cars. Intermodalism provides a system that allows people to travel more efficiently and cheaply.

L. **Transportation services and methods have led to a population that is regularly on the move.** For instance, people today can travel to foreign lands or to sites of interest hundreds of miles from home as quickly as they used to take a relatively short trip into town in a wagon 200 years ago.

M. **The design of intelligent and non-intelligent transportation systems depends on many processes and innovative techniques.** For example, the development of an intelligent transportation system — smart highways with electronic message boards, for instance — require the use of coordinated subsystems to determine capacity of lanes, traffic flow, and potential congestion problems. Non-intelligent transportation systems, such as walkways and bicycle paths, attract individuals and groups of people through innovative designs that capitalize on natural settings and provide convenience.

STANDARD

19 Students will develop an understanding of and be able to select and use manufacturing technologies.

Manufacturing is the production of physical goods. These goods can range from tools, such as kitchen appliances and computers to products, such as shoes and tennis balls.

The manufacturing of goods has changed tremendously over the last century, especially in recent years. Before manufactured goods became widely available, many goods were custom made — individuals made each product by hand, one at a time. With developments, such as standardized parts, assembly lines, and automation, production changed dramatically. For the first time, goods became cheaper as more of them were produced, an effect known as economies of scale. As machines became more accurate, making more complex items with interchangeable parts became possible. The first interchangeable items were handmade parts for firearms and plows produced before machine-based industry had developed. As previously discussed in Standard 18, some industries use a process called just-in-time (JIT) production, in which supplies, components, and materials are delivered just when they are needed. This process, which is designed to reduce the need to store inventory, places the burden on the suppliers to deliver quality parts and materials on an as-needed basis.

All goods are made of materials, and without these material resources, production is impossible. Although every material can be traced to one or more natural resources, very few materials can be used in their natural form. They must first be processed to some degree before they can be used to produce goods that are ready for market. For example, some clothing is made of cotton. But before cotton can be used to make clothing, it must first be harvested, processed, and woven into cloth. The same is true of all materials from steel and lumber to plastic. Materials must first be processed into standard stock, which in turn is used to make manufactured products. The processing of materials into standard stock is referred to as primary manufacturing.

We live in a global economy where products made in the United States, Finland, Japan, Taiwan, Malaysia, Mexico, Canada, or any other country are sold and used worldwide. Our lives have been enhanced because of manufacturing technology. It provides a segment of the population with jobs; it is a major factor in the economy (Gross National Product); and it provides us with many products that improve the quality of life.

GRADES

K-2

Manufactured goods include most of the items that a student will wear or bring to school, from jeans and backpacks to pencils and textbooks. Some goods last a long time, while others are designed to be used and thrown away. At the K-2 level, students will learn that a manufactured good is an item made for consumption or use. These goods are regularly being designed and redesigned to be made better, more cheaply, and more quickly as the technology advances. One way for students to see such improvement directly is by having them compare goods used today with similar ones from 10, 20, or even a 100 years ago. A visit to a museum or viewing a videotape on the historical development of manufacturing may provide resources for such a comparison.

Students also will begin to develop an understanding of how products are designed, produced, tested, packaged, and marketed. To reinforce this understanding, students could simulate this process by producing a snack comprised of nuts and dried fruit. They could begin by conducting a survey among their peers to learn what type of nuts and fruit were preferred for a snack mix. They should include questions about allergies — peanuts, for example. From the results of their testing, they could determine an appropriate recipe for the snack mix. Next, they could devise simple processes for controlling the amount of each ingredient that would go into each package. They could market the product to others, and finally, they could manufacture and package the snacks. In the process of learning how goods are made, students will have many opportunities to learn about teamwork and job specialization. They will study how people work and the ways they earn money.

In order to select, use, and understand manufacturing technologies, students in grades K-2 should learn that

A. **Manufacturing systems produce products in quantity.** Products can be made faster, cheaper, and better through the use of technology. People have different roles in the manufacturing process. If people work together, they can produce much more than if they work alone to make the same product.

B. **Manufactured products are designed.** Designers and engineers anticipate what people want and need with the intention that products will be bought. Some things are designed to be thrown away, while others are made to last a long time.

GRADES
3-5

In grades 3-5, students will gain a greater understanding of how goods are manufactured. They will explore how proper servicing of the goods ensures that they work properly and are up-to-date. As a result of experiences in designing, producing, and marketing a product, students will attain a greater insight into how to purchase products.

Everyone in our society uses manufactured goods on a daily basis. For example, students ride to school in a bus, wear clothes, and use pens and pencils. They have come to depend on these items. Starting at this early age, students should be taught how to be wise consumers. They should discuss how they make decisions, and whether those decisions are based on advertising, color, cost, warranties, peer pressure, quality, or a combination of these factors.

The use of technology and its impact on the environment should be considered in the design of goods. Designers of these items should take into consideration how long the product will last, what waste is produced, as well as what happens when the product is no longer in use and is discarded. Recycling is an important factor in the life of goods and structures.

Manufacturing systems generally include two steps. First, natural (raw) materials that are grown or extracted from the earth, are converted into standard stock items. Second, these standard stock items and some natural or synthetic materials are used to make products. For example, trees are made into lumber, and the lumber is then used to make furniture.

At this level, students should discuss and experiment with the various processes used in manufacturing systems. Some of the processes include designing the item, gathering inputs (e.g., materials and energy), using tools and machines to change the form of the materials, manufacturing and marketing the finished products. Chemical technology also can play an important role as a processing technology because it modifies and alters chemicals, elements, and compounds in order to produce materials with the desired chemical properties. Throughout the manufacturing process, feedback should be collected in order to monitor the quality of these products. Feedback should also be gathered after the item is completed to determine if consumers like the product and its effects on the family and society. These effects can be recognized by enhanced customer satisfaction, improved product sales, more jobs, and a stronger economy.

In order to select, use, and understand manufacturing technologies, students in grades 3-5 should learn that

C. **Processing systems convert natural materials into products.** Materials, which come directly from nature or are created by humans (synthetic), are essential inputs in the manufacturing system.

D. **Manufacturing processes include designing products, gathering resources, and using tools to separate, form, and combine materials in order to produce products.** Many manufactured products are composed of standardized parts, which reduces the cost of manufacturing and makes it easier to service and repair the products. It is important to consider the manufacturing process during the design of a product. It is also important to consider how long materials will last, what their effect will be on the environment, and how they will be disposed of.

E. **Manufacturing enterprises exist because of a consumption of goods.** When these enterprises produce goods that people need and want, they will spend money to purchase them. This cycle provides jobs and helps the economy.

GRADES
6-8

Students routinely use computers, book bags, bicycles, and watches. Many of them take these products for granted. Although they know how to buy them and, in many cases, how to use them, their understanding often goes no deeper. Students at this level will build upon their knowledge from prior grade levels to develop an in-depth understanding of where these products and systems come from, how they are made, how to use them appropriately, how they are marketed, and how to dispose of them. For these goods to continue to work properly and efficiently, they must be serviced. Services include those activities that provide support for a good after it has been sold or leased.

Students must be aware of how the manufacturing processes can have impacts on people and the environment. They should explore and experiment with various techniques for designing and developing processes and systems that are compatible with the natural environment.

Manufactured goods are classified according to their longevity — durable or non-durable, for example. Many of these goods are given a guarantee that protects the buyer for a specified period of time.

Manufacturing processes encompass the designing, producing, and marketing of goods. Some things are made one at a time, such as homemade clothes, custom cabinets, industry processing equipment, and some musical instruments. With the growth of modern manufacturing plants, however, this custom production has become relatively rare. With the use of machines, computer-aided design (CAD), automation, robots, and moving assembly lines, many identical items are produced very quickly, often with little intervention from humans.

Manufacturing processes include both mechanical and chemical processes. Students should have opportunities to experience processes that include separating, forming, combining, and conditioning materials. The students should understand that some materials must be obtained from the earth through such processes as harvesting, drilling, and mining. Many of these materials then are changed into standard stock materials before they are used to produce goods. For example, iron ore, limestone, and coke can be combined to make steel; steel can be processed into bars, rods, and pipes; and then these parts can be used to manufacture cars, for example. Middle-level students should test and evaluate various types of materials and processes before selecting the most appropriate ones to use when they are working on a product in the laboratory-classroom.

Products need to be marketed before they can be distributed and sold. Marketing involves researching potential customers and advertising the product. Servicing is important after a product is in use. In our country today, more people are employed in the service sectors of the economy than in the manufacturing sectors of the economy.

In order to select, use, and understand manufacturing technologies, students in grades 6-8 should learn that

F. Manufacturing systems use mechanical processes that change the form of materials through the processes of separating, forming, combining, and conditioning them.

Separating includes cutting, sawing, shearing, and tearing. Forming includes bending, shaping, stamping, and crushing. Combining includes gluing, welding, riveting, and using fasteners (e.g., nuts, bolts, and screws). Conditioning involves processing materials, such as by heating or cooling, to improve their structures. Tempering metals is an example of conditioning.

G. **Manufactured goods may be classified as durable and non-durable.** These classifications are based on the life expectancy of a product or system. Durable goods include automobiles, kitchen appliances, and power tools, while non-durable goods include toothbrushes, disposable diapers, and automobile tires. Manufactured goods have lifecycles, including initial planning and design, and continuing to their eventual disposal. Factors to be considered include what by-products were created, when the item was made, and how the item will be disposed of at the end of its life cycle.

H. **The manufacturing process includes the designing, development, making, and servicing of products and systems.** This process includes the use of materials (natural and synthetic), hand tools (e.g., hammers and scissors), human-operated machines (e.g., drills, sanders, and sewing machines), and automated machines (computer-controlled). Manufacturing systems have greatly increased the number of products available while improving quality and lowering costs. In general, machines, many of which are computer controlled, are capable of producing higher quality goods than an expert craftsperson could do individually. Services include those activities that provide support for a product or system after it is sold or leased. These services could include installing, troubleshooting, maintaining, and repairing.

I. **Chemical technologies are used to modify or alter chemical substances.** The products of chemical technologies include synthetic fibers, pharmaceuticals, plastics, and fuels.

J. **Materials must first be located before they can be extracted from the earth through such processes as harvesting, drilling, and mining.** Because few materials occur in nature in a usable state, they must be changed into new forms before they can be used as inputs in the manufacturing process. There also are other resources that are needed for manufacturing systems to operate properly, such as finances, people, tools and machines, information, and time. Natural (raw) materials are typically converted into standard stock items, which, in turn, become the resources that are used by manufacturers.

K. **Marketing a product involves informing the public about it as well as assisting in selling and distributing it.** Marketing entails assessing what the public wants and then advertising and selling products to the buyers.

VIGNETTE

A Team Approach to Plastics

This vignette presents some activities that deal with plastics as a manufactured product. Students not only study plastics, but they also design and make plastic products. Finally, they communicate to others the material which they learned. [This vignette highlights some elements of the Grades 6-8 *Technology Content Standards* that provide connections with Standards 3, 8, 9, 10, 11, 12, 15, and 19.]

The seventh-grade technology, language arts, and science classes worked together to implement an interdisciplinary unit on making and recycling plastics. The students were challenged to investigate the chemistry of plastics, the various products made with plastic, how new products are made from recycled plastic and plastic scraps, and the benefits the community receives through the use of plastics and recycling.

The students developed an action plan to complete the project, interviewed various engineers, scientists, technologists, and industry personnel, and toured a local plastic manufacturing plant and a recycling facility. In the course of the unit, the students worked with various types of plastics and designed and made examples of the individual objects they had investigated.

Students also conducted research on how synthetic materials differ from natural materials. Additionally, the teacher asked the students to create a company that involved the design, development, production-line operation, and assessment of a plastic product made in quantity. During their activities, the students documented their work by using videos and cameras. They produced a three-minute presentation describing what they had learned and then broadcast the segment on the school's television station. Finally, they produced a similar presentation for their school's World Wide Web site.

GRADES

9-12

Products and systems have a certain life expectancy. Many products are made with built-in obsolescence, which means that they are no longer used after an expected period of time. This trend has contributed to a throwaway mentality in our society. Many people regularly discard old products and buy new ones, which in turn has created large amounts of waste. Students should explore this trend and look for various ways to use products longer through proper maintenance and repair, as well as through recycling. They also should conduct research on the depletion of resources and develop ways to sustain them.

The basic processes in manufacturing can be classified into separating, forming, combining, and conditioning. To gain a deeper understanding of the manufacturing process, students can produce items on an assembly line. They should safely use various materials, tools, and processes in order to design, make, and assess their products. Materials have many qualities, and they can be classified by how they are found or made, such as natural materials (found in nature), synthetic materials (human made), and a mixture of both natural and synthetic materials. The study of chemicals is important because they are significant resources in today's world. In addition, students should be familiar with the work that people do in manufacturing and how products are marketed to consumers. Marketing and advertising products is important to better assure sales. Once the goods are sold or leased through a marketing effort, servicing becomes an important concern. One type of servicing is maintenance, which, if performed on a regular basis, can increase the life expectancy of goods. Students can learn how to service various products through such processes as installing, repairing, altering, maintaining, and upgrading.

Manufacturing systems can be classified according to the type of item being produced. Customized production involves making a single item that was designed with the needs of an individual in mind. Batch productions turn out parts or components — typically referred to as standard stock items — that are assembled at some later time. In developed countries, continuous (assembly-line) production is the most common way to make products today. One of the key features of assembly lines is the use of interchangeable parts, which is an important concept for students to learn about and experiment with at the high school level.

In order to select, use, and understand manufacturing technologies, students in grades 9-12 should learn that

L. **Servicing keeps products in good operating condition.** Servicing processes include installing, diagnosing and troubleshooting, recalling, maintaining, repairing, altering and upgrading, and retrofitting. Some products are designed for eventual obsolescence. Sometimes obsolescence is due to changing styles — the colors and shapes of kitchen appliances, for example.

M. **Materials have different qualities and may be classified as natural, synthetic, or mixed.** Examples of materials found in nature are wood, stone, and clay. Synthetic materials are human-made, such as plastics, glass,

GRADES
9-12

and steel. Mixed materials are a combination of natural and synthetic materials, such as plywood, paper, and wool-polyester blends of fabric.

N. **Durable goods are designed to operate for a long period of time, while non-durable goods are designed to operate for a short period of time.** Examples of durable goods are steel, furniture, and stoves. Non-durable goods, or consumable goods, include food, batteries, and paper.

O. **Manufacturing systems may be classified into types, such as customized production, batch production, and continuous production.** Customized production meets the specific needs and wants of an individual or small group by producing a single item or small quantities of goods. Batch production generates parts to be assembled later into larger products. Continuous production makes items on an assembly line or in a processing plant.

P. **The interchangeability of parts increases the effectiveness of manufacturing processes.** Components of a product or system must be interchangeable. Since manufacturing has become global, international standards for the interchangeability of parts have emerged.

Q. **Chemical technologies provide a means for humans to alter or modify materials and to produce chemical products.** Chemical technologies have been used to improve the health and well-being of humans, plants, and animals.

R. **Marketing involves establishing a product's identity, conducting research on its potential, advertising it, distributing it, and selling it.** Marketing should be considered from the design stage of a product to its final sale. Large corporations typically have their own marketing departments, whereas smaller companies with limited resources may contract with a marketing firm.

STANDARD

20 Students will develop an understanding of and be able to select and use construction technologies.

Humans have been building structures for millennia. The Chinese erected the Great Wall; the Egyptians built pyramids; the Greeks constructed elaborate buildings; and the Romans created remarkable roads. Today, many of the same principles for building structures used centuries ago are still being applied. Large structures, for instance, need substantial foundations, and for many centuries, builders have known that triangles have more compressive strength than rectangles for making roofs, bridges, and high-rise buildings. Hard materials (steel and concrete) have also been found to withstand some weather conditions better than softer materials (wood and limestone).

The processes involved in designing and making structures are typically referred to as construction. People from many different professions work in the construction industry, including architects and engineers, builders, estimators and bidders, carpenters, plumbers, concrete workers, and electricians.

Structures, such as houses, office buildings, agricultural-storage facilities, roads, and bridges, serve a variety of purposes. In some cases, they are designed primarily to provide shelter and a place to live. Other structures are used for entertainment and recreation, such as concert halls, amusement parks, and football stadiums, and yet others are primarily for work, such as factories and oil drilling rigs. Another major class of structures includes those that support transportation, such as bridges, roads, and airplane hangers.

Some structures are temporary, while others are permanent. Such structures as scaffolding, cofferdams (a temporary structure used to create a dry space in water so that a pier or bridge foundation can be built), and even tree houses are deliberately designed to last only for brief periods at a given location. As a result, less time and expense are invested in the construction. Permanent structures are those that are designed and constructed to last for a long time. Examples include parking garages, office buildings, water towers, school buildings, bridges, and air-traffic control towers. Even permanent structures, however, will eventually wear out or become obsolete.

Whereas manufacturing typically uses assembly-line processes, construction typically uses customized processes. Even though many houses and office buildings look alike, they are usually built one at a time, and each one tends to have defining characteristics. Many structures have unique, "one of a kind" designs. Another difference between manufacturing and construction is that manufacturing usually takes place in factories, whereas construction typically occurs on a building site.

GRADES
K-2

At an early age, children develop ownership of "their place," typically a room in an apartment or house. This place is a part of the constructed environment in which they can receive protection, shelter, and comfort. During the early grades, students can expand their understanding of the constructed environment to include the places with which they interact on a daily basis. In addition to their homes, such environments can include their school, a library, a church, stores, and places where parents work.

The construction of shelter for human protection has evolved from caves and huts to houses, apartment complexes, and office buildings. Technological advances in such areas as glass windows, tile making, lighting, furniture, air conditioning, and electrical utilities have helped improve the conveniences and comforts of dwellings.

Almost every community has an active construction site — homes, buildings, sidewalks, bridges, or roads that are being built or refurbished. At these community examples, children can actually see the construction process in progress. They can see that special materials, such as concrete, bricks, lumber, steel, and glass are being used. They also can observe the many different people who are involved in the construction process. In addition, students should be given the opportunity to design and fabricate models of construction works in the laboratory-classroom — involving them in making a planned model community, for example. Some students could be assigned to work on roads, others could design buildings, while others could work with utilities and landscapes.

In order to select, use, and understand construction technologies, students in grades K-2 should learn that

A. **People live, work, and go to school in buildings, which are of different types: houses, apartments, office buildings, and schools.** Buildings are designed, built, and maintained by people. Special materials are used to make buildings. Historically people tended to use materials available in their communities for building materials. With the advent of modern ways to convert natural materials into building materials and improved transportation systems, special materials are now available, including lumber, stone, brick, and plywood.

B. **The type of structure determines how the parts are put together.** The way the parts are arranged or put together to form a whole determines the type of structure. Some common structures include buildings, which protect people and goods, and roads and bridges, which support transportation.

GRADES

3-5

Students at this level should begin to understand the notion of community development. They live in a residential area, go shopping at local stores, and use local parks. Students should understand that these constructed areas were planned and designed.

As with other technologies in the designed world, resources are needed as inputs into the construction process. These resources include tools and machines, materials, information, energy, capital (money), time, and people. Maintenance is an important concept in the preservation of buildings. People, including children, can cause wear and tear on such things as buildings, roads, and bridges. Weather also contributes to deterioration, and regular maintenance is important for structures to last.

By the time students complete the fifth grade, they will have enlarged their concept of the constructed environment to encompass more than buildings. The constructed environment also includes trails, roads, railways, utility services, dams, pipelines, waterways, airports, and bridges, which allow people to move freely about the community and enable goods to be moved from place to place.

Students should have the opportunity to design and build models of structures. This process can provide a meaningful way for them to develop spatial relationships. They also should begin to recognize that many systems are used in structures that provide such conveniences as drinking fountains, toilets, lights, and comfortable temperatures, which make their lives more enjoyable.

In order to select, use, and understand construction technologies, students in grades 3-5 should learn that

C. **Modern communities are usually planned according to guidelines.** Special areas are designated for schools, stores, parks, houses, apartments, manufacturing plants, and offices. Sidewalks, trails, roads, and bridges provide routes for people to move throughout the community. In addition to building materials — sand, gravel, lumber, and brick — specialized tools and machines and large amounts of money — are needed in the construction industry as well as time, energy, land, and people.

D. **Structures need to be maintained.** Weather and usage cause deterioration in any structure.

E. **Many systems are used in buildings.** Some are simple, while others are complex. For example, a plumbing system provides water and eliminates sewage, and a heating and cooling system maintains comfortable temperatures in summer and winter. Other technologies are an integral part of a building as well. For example, the telephone is a part of communications technology. When building a house or office building, one part of the whole process is installing telephone lines so that the people who live or work in that structure can communicate with the outside world.

GRADES

6-8

It is important for students in the middle grades to become involved not only in designing and making models of structures, but also in developing an understanding of the importance of the constructed environment in their daily lives. Students will learn through activities in the laboratory-classroom about types of structures and the purposes that each serves, the importance of proper design, the importance of maintaining structures, the use of subsystems in a building, and the need for community planning, including laws and regulations.

Through these activities, students will understand that the foundation of a structure is built to provide the footing or underpinning on which it stands. The foundation provides a stable base, which is level and solid.

The constructed environment is a complex array of structures which are used for many different purposes and which have been constructed over a long period of time. It is an environment that is undergoing continual change. Some structures can fall into disrepair, and the original design no longer meets current needs. As a result, many structures are demolished, and in their place, new structures are built. An understanding of how and why these changes take place will help students understand the world in which they live.

Students should have opportunities to design, use, and assess buildings and materials. They should understand that buildings have subsystems that are used to do specific things. For example, the electrical system is used to light the building, and a heating and air conditioning system provides comfortable temperatures. There are many types of materials needed in a construction job that are used to provide form, decoration, protection, and strength. Materials can be natural (rocks, timber) or synthetic (bricks, asphalt, concrete, steel).

In order to select, use, and understand construction technologies, students in grades 6-8 should learn that

F. **The selection of designs for structures is based on factors such as building laws and codes, style, convenience, cost, climate, and function.** Building laws and codes are part of the city or county regulations for construction.

G. **Structures rest on a foundation.** The structures determine the type of foundation needed. Foundations can be made from such materials as concrete, steel, and wooden poles.

H. **Some structures are temporary, while others are permanent.** Many times, temporary structures are built to aid the construction of permanent structures. For example, scaffolding is often assembled to support workers who lay bricks, and forms are used as containers to hold poured concrete. There are many different types of interior and exterior building materials. These materials include brick, rock, stone, siding, log, wood, brick veneer, plywood, metal, wallboard, concrete, glass, and straw and mud.

I. **Buildings generally contain a variety of subsystems.** These subsystems include waste disposal, water, electrical, structural, climate control, and communication. Most of these subsystems are referred to as utilities.

GRADES

9-12

By the time students graduate from high school, they should know about a number of factors associated with evaluating and purchasing structures of various types. Virtually all citizens are affected in one way or another by construction technologies. They purchase and live in homes. They work in offices and factories. They receive radio and telephone signals that have been transmitted through towers. They drive over bridges and park in multi-deck garages.

Structures are explored in greater depth at this level. Students should design structures and make models of them. They should understand that certain structures can be thought of as part of a much larger system that underlies the functioning of the entire society. Roads and bridges, airports and railways, electrical transmission and distribution systems, dams, ships, water-treatment plants, water-supply systems, and sewers all constitute the physical infrastructure of a society. An adequate infrastructure is necessary for other technologies to function efficiently.

At the high school level, students should be able to identify the various materials and systems that comprise buildings. These include utilities, such as water, waste, electrical, climate control, telephone, and gas, as well as component systems, such as foundations, framing, insulation, and lighting.

Since a home is often the single largest financial investment an individual makes, it is important that citizens be equipped to assess the quality of homes and other structures, including the quality of processes and materials used in the construction job. At the very least, they should know whom to contact for professional assessments and have sufficient knowledge to interpret inspection reports.

A number of factors are used to guide the process of designing and making structures. Students should understand that various requirements are used to make construction decisions. Some relate to personal preference, such as location, style, and size. Other factors deal with legal restrictions, such as zoning laws, building codes, and professional standards. Additionally, the selection of requirements often depend on the kind of structure. For example, a primary consideration for a bridge is strength, whereas style and affordability are important criteria for many homes.

Periodic improvement or even renovation of a structure is vital to extend its lifetime or improve its usefulness. In urban areas, two-lane highways are widened to four lanes to accommodate more traffic, for instance. A structure can be altered to change its size, appearance, or function. In some instances, buildings are torn down in order to make room for new ones. Students should realize that as with other technologies, decisions related to construction have impacts on individuals, society, and the environment. An important purpose of construction is to provide shelter and structures for humans.

In order to select, use, and understand construction technologies, students in grades 9-12 should learn that

J. **Infrastructure is the underlying base or basic framework of a system.** An infrastructure often includes the basic buildings, services, and installations needed in order for a society or

GRADES
9-12

government to function, such as transportation, communication, water, energy, and public information systems.

K. **Structures are constructed using a variety of processes and procedures.** In some cases, the procedure used depends on the type of material available. For example, welds, bolts, and rivets are used to assemble metal framing materials. Sometimes procedures are selected as a function of cost, skills, and preference of the worker, or the level of quality desired. Citizens should be equipped to evaluate the appropriateness of procedures used.

L. **The design of structures includes a number of requirements.** One of the most important design constraints with structures is function. For example, the function of houses is to provide safe and pleasant shelter for families, whereas the primary function of a bridge is to carry loads over barriers or obstructions. Other important constraints include appearance, strength, longevity, maintenance, and available utilities. The design and construction of structures are regulated by laws, codes, and professional standards. Common design constraints used by engineers and architects in the design of structures include style, convenience, safety, and efficiency.

M. **Structures require maintenance, alteration, or renovation periodically to improve them or to alter their intended use.** Structures must be designed and constructed to provide for maintenance. Most structures are comprised of a variety of systems, each of which commonly requires maintenance. For example, because electrical and telephone systems typically need to be upgraded in office buildings, easy access must be included in the original design process — renovating a hotel to serve as a nursing home, for example. Sometimes, alterations and renovations are necessary because a structure has become outdated or in need of repair.

N. **Structures can include prefabricated materials.** Certain kinds of materials are appropriate for some prefabricated structures and parts of structures, while others are not. For example, for various reasons, wood, concrete, and steel are commonly used as prefabricated frames for houses, bridges, and buildings. One important quality variable concerns the type and quality of materials used and the support loads required. Prefabricated sections of buildings can be set in place to reduce costs, and a wide range of options are typically available at different costs.

7

VIGNETTE

A Look at Energy Efficient Homes

This vignette presents an activity to design and construct a model home. Criteria and constraints are given to guide the students in their problem-solving processes. [This vignette highlights some elements of the Grades 9-12 *Technology Content Standards* which provide connections with Standards 8, 9, 10, 11, 12, 13, 16, and 20.]

The city of Westlake and the surrounding areas experienced an accelerated growth in the construction industry, especially in new home construction. The local high school technology teacher, Mr. S, thought it would be helpful for his students, as future consumers, to have an in-depth understanding of the housing industry and to know about the latest developments in home construction techniques, materials, and practices.

Mr. S decided to organize a lesson where students were invited to participate in designing an energy-efficient home for a family of four. He guided the students to consider all forms of energy and not to limit their imaginations. Students were instructed to consider costs of using energy-efficient designs and how those costs might affect the resale value of a home.

The students in the technology classes were challenged to design, draw, and build a scale model of a residential home using heating and cooling systems that were energy efficient, aesthetically pleasing, functional, marketable, and innovative. The house also had to accommodate a family of four with a maximum size of 2100 square feet. The students had to work within a budget of $150,000, and they had nine weeks to complete the project.

The students began by researching homes in their area that already incorporated features that were required in their home. They conducted library and Internet searches to learn about the latest materials and techniques available in the housing industry. Students also interviewed local architects and building contractors to learn about various practices and how they were integrating innovative features. For example, they learned about incorporating increased day lighting, which takes into account the home's orientation, into the design of the home. They also learned about designing and installing environmentally sound and energy-efficient systems and incorporating whole-home systems that are designed to provide maintenance, security, and indoor-air-quality management.

The students then began the process of sketching their homes. Many students had to gather additional research as they realized they needed more information to complete their sketches. Using their sketches, the students built scale models of their homes out of mat board.

A group of building industry professionals from across the area was invited to evaluate students' work and provide feedback on their ideas in several categories, including design, planning and innovations, energy conservation features, drawing presentation, model presentation, and exterior design.

As a result of this experience, the students learned firsthand what it takes to design a home for the 21st century. Students also learned how to successfully plan and select the best possible solution from a variety of design ideas in order to meet criteria and constraints, as well as how to communicate their results using graphic means and three-dimensional models.

8 Call to Action

Because technological literacy is so important to all of us, ITEA is calling for interested parties to join it in advancing the cause of technological literacy as laid out in these standards. Foremost, ITEA encourages the adoption of *Technology Content Standards* in states, provinces, and localities.

8 Call to Action

In order to meet the goal of technological literacy for all, a collaborative effort among interested parties — teachers, principals, superintendents, supervisors, teacher educators, students, parents, educational equipment providers and publishers, engineers, scientists, mathematicians, technologists, and the community at large — will be essential.

In conjunction with the nationwide adoption of these standards, ITEA recommends that the following topics or groups be addressed:

- Curriculum Development and Revision
- Learning Environments, Instructional Materials, Textbooks, and Other Materials
- Technology Education Profession
- Students
- Overall Education Community
- Parents and the Community
- Engineering Profession
- Other Technology Professionals
- Business and Industry
- Researchers
- Additional Standards

Curriculum Development and Revision

Technology Content Standards defines what the study of technology in grades K-12 should be, but it does not lay out a curriculum — that is, it does not specify how the content should be structured, sequenced, and organized. This task is left — as it should be — to individual teachers and other curriculum developers in the schools, school districts, and states and provinces. Because the 20 standards with their supporting benchmarks contained in *Technology Content Standards* outline what a curriculum should accomplish, developers of new and existing curricula are encouraged to use this document as a roadmap. After a curriculum is devised, the next step is to convert it into day-by-day lesson plans used by teachers in their laboratories-classrooms.

To make it easier for educators to implement the standards, ITEA's Center to Advance the Teaching of Technology & Science (CATTS) will support localities, states, and provinces in developing curricula based on *Technology Content Standards*. In addition, ITEA will provide instructional materials, publications, and professional development activities to assist teachers in putting the standards into practice.

Technology Content Standards defines what the study of technology in grades K-12 should be, but it does not layout a curriculum — that is, it does not specify how the content should be structured, sequenced, and organized.

Learning Environments, Instructional Materials, Textbooks, and Other Materials

If the study of technology is to be effective, the facilities, equipment, materials, and other parts of the infrastructure must be appropriate and current. In particular, instructional materials and textbooks used in the study of technology should be modified to reflect *Technology Content Standards*. The suitability of instructional materials, modules, and textbooks can be assessed by comparing their content to the content of this document. Similarly, it is recommended that the developers and publishers of instructional material follow the "Administrator's Guidelines For Resources Based on *Technology Content Standards*" presented in Chapter 2.

Technology Education Profession

Those who educate technology teachers should review and revise undergraduate and graduate degree programs by using Technology Content Standards as the basis for teaching technology.

Support from the technology education profession is vital to the acceptance and implementation of *Technology Content Standards*. By using this document as a basis for modifying their instruction, teachers will demonstrate the importance of technological studies, the value of technological literacy, and their own abilities to teach about technology.

In-service programs must be developed to teach technology educators how to implement *Technology Content Standards*. Supervisors are encouraged to provide support and philosophical leadership for reform in the field because they are in an ideal position to implement long-range plans for improving the delivery of technology education subject matter at the local, district, state, and province levels.

Those who educate technology teachers should review and revise undergraduate and graduate degree programs by using *Technology Content Standards* as the basis for teaching technology. Furthermore, strategies can be designed and implemented for recruiting and preparing a sufficient number of newly trained and credentialed technology education teachers. Alternate certification programs may be established in states and provinces with serious shortages of technology teachers.

Students

The Technology Student Association (TSA) provides co-curricular and extra-curricular educational experiences that enrich students' learning about technology. To further that goal, TSA is encouraged to use *Technology Content Standards* in the development of new activities and competitive events.

The Junior Engineering Technical Society (JETS) also has a number of services and student activities that could be enhanced by incorporating the standards and benchmarks from *Technology Content Standards*. On the collegiate level, the Technology Education Collegiate Association (TECA), a university-based student organization for pre-service teachers, can incorporate *Technology Content Standards* in its programs.

Since there is a major demand for technology teachers, students who have an interest in both teaching and technology are encouraged to consider becoming technology teachers. They can choose from a number of universities that offer pre-service technology teacher education programs that license teachers.

Everyone interested in technological literacy is encouraged to understand the importance of the study of technology and support it as a basic field of study in public schools by developing policies and providing funding that will allow for the implementation of Technology Content Standards *in kindergarten through twelfth grade.*

Overall Educational Community

At the elementary level, the regular classroom is where technology should be taught. Although elementary teachers may initially feel unqualified to teach technology, experience has shown that with appropriate in-service training they perform exceptionally well and excel at integrating technological concepts across the curriculum. For the study of technology to become an integral part of elementary curricula, it is recommended that all elementary teachers have courses in technology in their undergraduate teacher preparation program in colleges or universities. This means that teacher preparation institutions are encouraged to include technology teacher education as a part of elementary teachers' undergraduate degree requirements.

At the secondary school level, teachers from other fields of study can help by becoming familiar with *Technology Content Standards.* If teachers of subjects other than technology apply *Technology Content Standards* in their own classes, students in middle and high school will learn about the rich interdisciplinary relationships between technology and other fields of study, such as science, mathematics, social studies, language arts and the humanities. Because of the particularly close interrelationship among technology, science, and mathematics, teachers are encouraged to work cooperatively in planning and implementing curricula that are based on standards from all three fields of study.

At the community college level, faculty and administrators who work with associate degree engineering technology programs are encouraged to become familiar with *Technology Content Standards.* This is important because high school graduates who are technologically literate may have an interest in pursuing careers in engineering technology. Additionally, community college personnel who work with tech prep programs in high schools with 2+2 configurations are asked to become familiar with *Technology Content Standards* so that articulated technology programs can be developed for those students who wish to pursue technical or technology related careers.

Beyond the classroom, school administrators — principals, curriculum developers, directors of instruction, superintendents, and others — can recognize the importance of technological literacy for all students and support the study of technology. To that end, they can provide the support and funding for the materials, equipment, and laboratories needed for the teaching of technology. Current staff should be provided with professional development activities and in-service programs that will prepare them to put *Technology Content Standards* into practice.

It is important that local school boards, state and provincial legislators, and public officials also become familiar with *Technology Content Standards.* They are encouraged to understand the importance of the study of technology and support it as a basic field of study in public schools by developing policies and providing funding that will allow for the implementation of *Technology Content Standards* in kindergarten through twelfth grade.

Parents and the Community

By supporting and reinforcing the concepts learned in school, parents and other caregivers will play a central role in the education of their children. Students' attitudes toward the study of technology will, in large measure, reflect the attitudes of their parents. Therefore, parents with positive stances towards the study of technology will impart those attitudes to their children.

Parents and other caregivers are in an ideal position to advance technological literacy for their children. Relatively few, however, will have been exposed to technology education during their school years. For this reason, many parents may have the wrong idea about what the study of technology involves. A common misunderstanding, for example, is confusing technology education with computer education or with educational technology (equipment and software used in laboratory-classrooms to enhance the teaching and learning process). To promote technological literacy for all students, parents are encouraged to embrace and understand the value of the study of technology.

Ideally, other members of the community can also become knowledgeable about the study of technology and the importance of standards as a means to bring about reform in education and promote technological literacy. Community leaders also need to help the youth in their communities by supporting quality technology programs.

Engineering Profession

By serving as champions of technological literacy for all, members of the engineering community will not only benefit society in general, they will also benefit their own profession. As noted by the National Research Council in the publication *Engineering Education: Designing an Adaptive System* (1995), the health of the engineering profession is dependent upon a wide range of factors:

> The nation's engineering education system includes not just higher education but also K-12, community colleges, and continuous (lifelong) engineering education. These elements are embedded in the larger society, whose political and economic influences typically affect engineering schools through the academic institution of which they are a part. Those socioeconomic and political factors also drive demand for engineers, as well as the supply, recruitment, and retention of engineering students. (p.40)

Technological literacy will benefit the engineering profession in a number of ways. As more students receive high-quality instruction in technology, for example, more will be likely to select engineering as a career. In the long run, improved engineering will strengthen the technological base of the economy and of society.

The engineering community can encourage technological literacy in a number of important ways. Those in charge of engineering programs at colleges and universities will advance technological literacy by supporting undergraduate and graduate degree programs in technology teacher education. To that end, college and university engineering faculty can collaborate with technology teacher education faculty in interdisciplinary courses. Some engineering institutions also may conduct pre-service summer schools for college students with majors in science or mathematics education. Some institutions might establish programs for engineering graduates who are interested

In particular, research is needed that explores the specific ways in which the study of technology enhances a student's education.

in teaching grades K-12, for example, by having engineering graduates work as visiting *per diem* teachers who receive credit toward a teaching certificate.

Through statewide consortia, engineering institutions could set up centers where K-12 teachers would acquire in-service training on teaching tools and topics in technology. Engineering colleges and universities might set up programs to "adopt" elementary and secondary schools and assist them in developing laboratory projects and classroom activities. Similarly, members of engineering societies might form partnerships with K-12 teachers to provide students with hands-on engineering experiences.

The engineering education community, perhaps through the National Academy of Engineering or the National Research Council, is encouraged to support ongoing efforts to reform K-12 science, mathematics, and technology at the national, state, provincial, and local levels. A task force may be established to examine the college curricula of students who are planning to teach K-12 mathematics, science, and technology. The task force could focus particularly on the technological literacy of these students and on what they are taught about engineering and engineering achievements. An effort can be initiated to redesign many undergraduate science courses for K-5 teachers to include more content about technology.

Other Technology Professionals

In addition to engineers, many other professionals support technological literacy for all students. Examples of some of these technology-oriented professionals include architects, computer scientists and programmers, industrial designers, technicians, draftspersons, equipment maintenance personnel, and others. They are encouraged to read *Technology Content Standards* and support its implementation in grades K-12 in their community and schools. Because we live in a technologically oriented world, collaborative efforts will be most beneficial to future generations.

Business and Industry

It is vital that business and industry leaders at the national, regional, and local level become more involved in school programs in general, and in particular, with technology programs. These individuals typically have both the resources and expertise to help implement *Technology Content Standards*. They are encouraged to become familiar with *Technology Content Standards* and work with local, state, and province personnel to improve technology programs by using the document as a guideline. Business and industry leaders also are encouraged to donate instructional materials and equipment to K-12 schools and to persuade professionals within their companies to join with K-12 teachers in providing relevant, hands-on experiences for students.

Recognizing the value of support from the business community, technology educators need to actively pursue backing for their programs from business and industry in their areas.

Researchers

Because few studies have examined K-12 technology programs, there is an acute need for additional research about technology. In particular, research is needed that explores the specific ways in which the study of technology enhances a student's education. This information will be important in

When we consider that many of the most influential people of the last millennium were inventors or innovators, the central role of technology is undeniable.

assuring decision-makers of the value of adding technology to an already overcrowded curriculum. Furthermore, research is needed to move *Technology Content Standards* forward and to provide support and direction for future revisions, as well as new standards in other areas.

Members of the technology education profession in particular, and the educational community in general, are encouraged to set forth new priorities for identifying a research agenda. This research agenda can be pursued by all members of the educational community, including teachers, local school administrators, and higher education faculty. Without an accepted and refined research agenda, future efforts could be haphazard and disjointed. The time has arrived for education professionals to realize that much more research must be conducted if the teaching and learning of technology is to advance. Research will be invaluable in developing future editions of the standards.

It is recommended that future international comparisons of students' achievements in the study of technology complement and add to those already being undertaken, such as the Third International Mathematics and Science Study (TIMSS). Additionally, research groups can be organized to study the relationships among technology, science, and mathematics.

Improvement of technological literacy begins with the implementation of these standards.

Additional Standards

Now that *Technology Content Standards* has been published, there is a need for developing further technology standards: assessment standards, program standards, and professional development standards (in-service and pre-service). Teachers and administrators are asked to look at their assessment techniques currently in place. They are encouraged to develop new curricula based on *Technology Content Standards* and to incorporate up-to-date assessment strategies for how well students meet these standards.

Technology Content Standards will be revised and updated periodically as more research is conducted, more experiences are documented, and more input is received.

Concluding Comments

When we consider that many of the most influential people of the last millennium were inventors or innovators, the central role of technology is undeniable. Consider the power and promise of some of their momentous technologies — Gutenberg's movable printing press, Galileo's telescope, Da Vinci's flying machine, Ford's Model-T, Edison's light bulb, and Brattain, Bardeen, and Shockley's transistor. Without them, the history of humankind would be vastly different. In light of the past, technological literacy for all students is a noble goal for the future.

Technology Content Standards does not represent an end, but a beginning. In other fields of study, the development of standards has often proved to be the easiest step in a long, arduous process of educational reform. Getting these technology standards accepted and implemented in grades K-12 of every school will certainly be far more challenging than developing them. This document, which is a starting point for action within schools, districts, states, and provinces, is aimed at making the study of technology essential for all students. Improvement of technological literacy begins with the implementation of these standards.

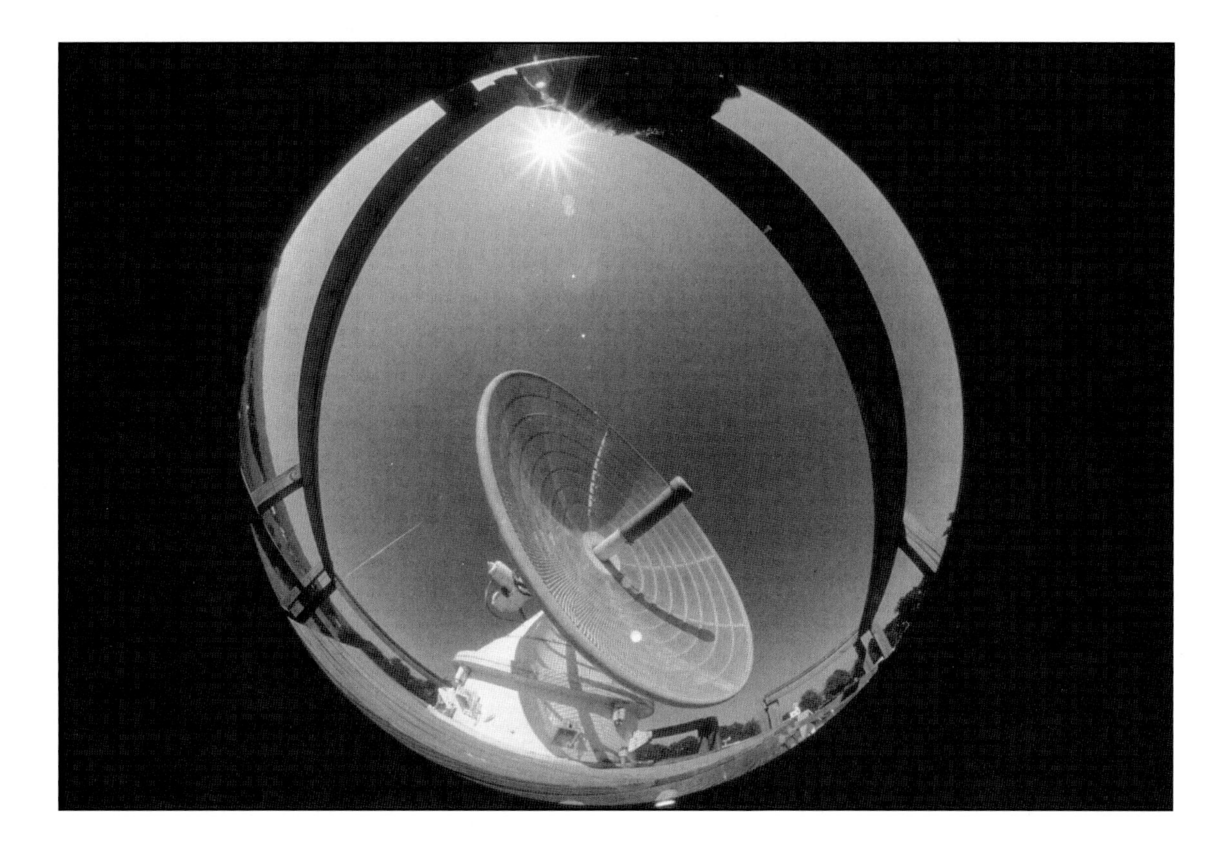

APPENDICES

History of the Technology for All Americans Project

ITEA, through its Technology for All Americans Project (TfAAP), published *Technology for All Americans: A Rationale and Structure for the Study of Technology* (*Rationale and Structure*) in 1996. This document provided the foundation for *Technology Content Standards* and established the guidelines for what each person should know and be able to do in order to be technologically literate.

The goal of *Technology Content Standards* was to build upon this work and to present a general-content framework for technology education. TfAAP created two advisory groups to assist with the development and refinement of the standards. One of these groups was the Standards Team, which was instrumental in advising the project and creating and refining the standards. The project also created an Advisory Group, which advised the project on the process of developing standards and gave specific input into the wording of the standards. Members of both of these committees, along with other groups and individuals that were instrumental in the development of *Technology Content Standards,* are listed in the acknowledgements.

In the development, consensus-building, and validation processes, six drafts of the document were generated. The following TfAAP chronology highlights key dates in this process.

FALL 1994 TO FALL 1996

- Phase I—Development of *Rationale and Structure.*

FALL 1996 TO SUMMER 1997

- Start of Phase II of the Technology for All Americans Project.
- The Standards Team began the process of developing the core of the standards.

SUMMER 1997 TO FALL 1997

- Draft 1 of *Technology Content Standards* was developed and distributed by mail, at hearings, and on the World Wide Web (WWW). Each individual or group had the opportunity to comment on the draft.

WINTER 1997 TO SUMMER 1998

- Based on the input received, the document was revised, and Draft 2 was produced. This draft focused on collecting input on the K-12 content standards only. Again, input was received through standards hearings, mail review, and the Internet on the project's web site.
- Draft 3 was produced from the winter of 1997 to the summer of 1998.

SUMMER 1998

- Field review of Draft 3 of *Technology Content Standards* by classroom teachers and administrators.

FALL 1998

- Draft 3 was finalized, mailed out for review, and additional hearings were conducted.

WINTER 1999

- Based on input received, ITEA and the project staff decided that the document should be revised again before being published.

SPRING 1999 TO FALL 1999

- National Research Council's Standards Review Committee (SRC) was formed and charged with reviewing the structure and format of the document.
- Draft 4 was developed and then reviewed by SRC and a Technical Review Committee (TRC) in August 1999.

FALL 1999

- Draft 5 of *Technology Content Standards* was developed and reviewed by SRC and a National Academy of Engineering (NAE) committee.
- Draft 6 was developed and reviewed by the NRC/SRC and the NAE Committees in late fall of 1999.
- Final layout and editing of *Technology Content Standards.*

WINTER TO SPRING 2000

- *Technology Content Standards* was published and disseminated.

APPENDIX

B Listing of The Technology Content Standards

THE NATURE OF TECHNOLOGY

STANDARDS

1. Students will develop an understanding of the characteristics and scope of technology.
2. Students will develop an understanding of the core concepts of technology.
3. Students will develop an understanding of the relationships among technologies and the connections between technology and other fields of study.

TECHNOLOGY AND SOCIETY

4. Students will develop an understanding of the cultural, social, economic, and political effects of technology.
5. Students will develop an understanding of the effects of technology on the environment.
6. Students will develop an understanding of the role of society in the development and use of technology.
7. Students will develop an understanding of the influence of technology on history.

DESIGN

8. Students will develop an understanding of the attributes of design.
9. Students will develop an understanding of engineering design.
10. Students will develop an understanding of the role of troubleshooting, research and development, invention and innovation, and experimentation in problem solving.

ABILITIES FOR A TECHNOLOGICAL WORLD

STANDARDS

11. Students will develop the abilities to apply the design process.
12. Students will develop the abilities to use and maintain technological products and systems.
13. Students will develop the abilities to assess the impact of products and systems.

THE DESIGNED WORLD

14. Students will develop an understanding of and be able to select and use medical technologies.
15. Students will develop an understanding of and be able to select and use agricultural and related biotechnologies.
16. Students will develop an understanding of and be able to select and use energy and power technologies.
17. Students will develop an understanding of and be able to select and use information and communication technologies.
18. Students will develop an understanding of and be able to select and use transportation technologies.
19. Students will develop an understanding of and be able to select and use manufacturing technologies.
20. Students will develop an understanding of and be able to select and use construction technologies.

C Compendium

Compendium of Major Topics for *Technology Content Standards*

Standards	Benchmark Topics Grades K-2	Benchmark Topics Grades 3-5	Benchmark Topics Grades 6-8	Benchmark Topics Grades 9-12
CHAPTER 3 NATURE OF TECHNOLOGY				
1 The Characteristics and Scope of Technology	• Natural world and human-made world • People and technology	• Things found in nature and in the human-made world • Tools, materials, and skills • Creative thinking	• Usefulness of technology • Development of technology • Human creativity and motivation • Product demand	• Nature of technology • Rate of technological diffusion • Goal-directed research • Commercialization of technology
2 The Core Concepts of Technology	• Systems • Resources • Processes	• Systems • Resources • Requirements • Processes	• Systems • Resources • Requirements • Trade-offs • Processes • Controls	• Systems • Resources • Requirements • Optimization and Trade-offs • Processes • Controls
3 The Relationships Among Technologies and the Connections Between Technology and Other Fields	• Connections between technology and other subjects	• Technologies integrated • Relationships between technology and other fields of study	• Interaction of systems • Interrelation of technological environments • Knowledge from other fields of study and technology	• Technology transfer • Innovation and Invention • Knowledge protection and patents • Technological knowledge and advances of science and mathematics and vice versa
CHAPTER 4 TECHNOLOGY AND SOCIETY				
4 The Cultural, Social, Economic, and Political Effects of Technology	• Helpful or harmful	• Good and bad effects • Unintended consequences	• Attitudes toward development and use • Impacts and consequences • Ethical issues • Influences on economy, politics, and culture	• Rapid or gradual changes • Trade-offs and effects • Ethical implications • Cultural, social, economic, and political changes
5 The Effects of Technology on the Environment	• Reuse and/or recycling of materials	• Recycling and disposal of waste • Affects environment in good and bad ways	• Management of waste • Technologies repair damage • Environmental vs. economic concerns	• Conservation • Reduce resource use • Monitor environment • Alignment of natural and technological processes • Reduce negative consequences of technology • Decisions and trade-offs
6 The Role of Society in the Development and Use of Technology	• Needs and wants of individuals	• Changing needs and wants • Expansion or limitation of development	• Development driven by demands, values, and interests • Inventions and innovations • Social and cultural priorities • Acceptance and use of products and systems	• Different cultures and technologies • Development decisions • Factors affecting designs and demands of technologies

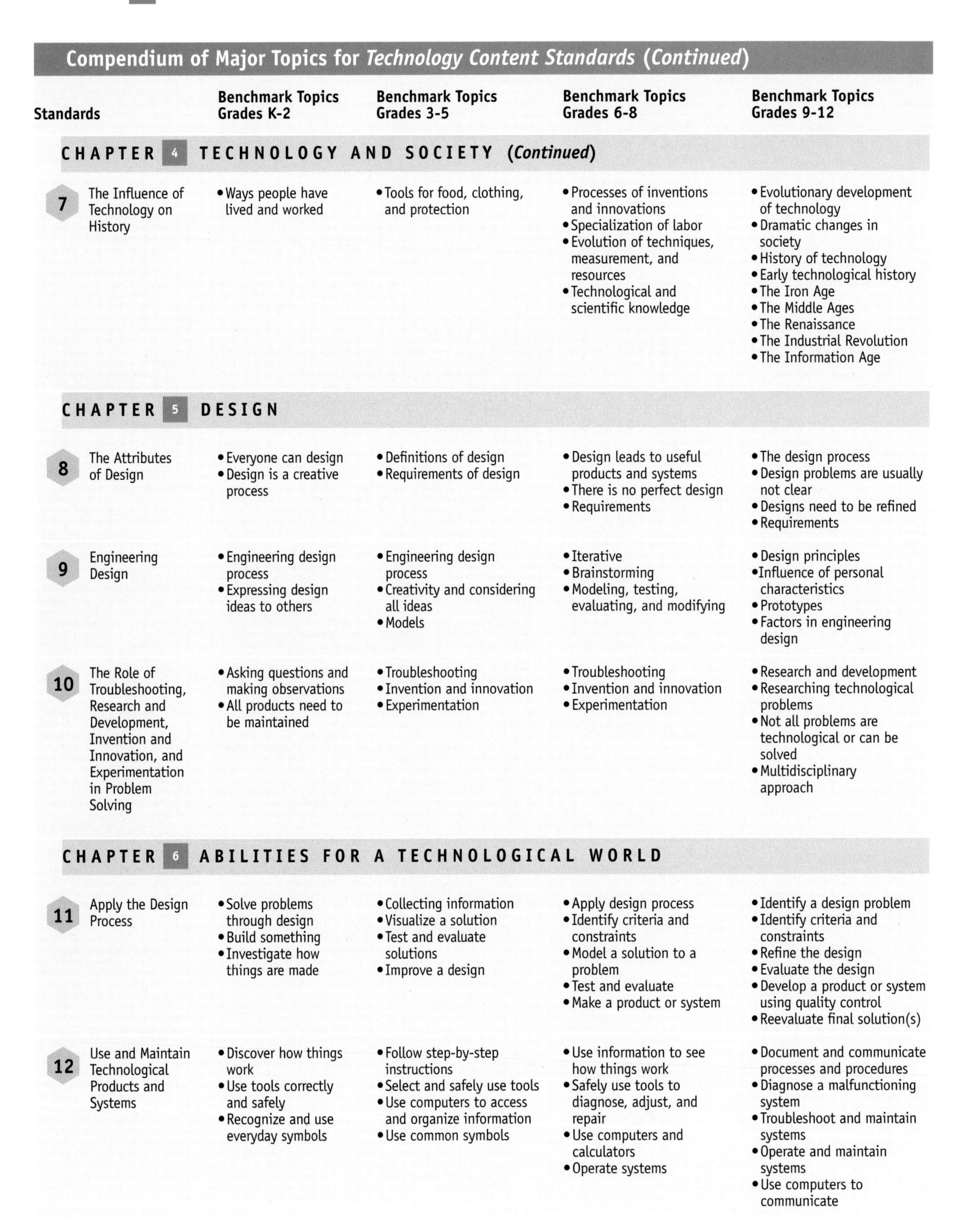

Compendium of Major Topics for *Technology Content Standards* (Continued)

Standards	Benchmark Topics Grades K-2	Benchmark Topics Grades 3-5	Benchmark Topics Grades 6-8	Benchmark Topics Grades 9-12
CHAPTER 4 TECHNOLOGY AND SOCIETY (*Continued*)				
7 The Influence of Technology on History	• Ways people have lived and worked	• Tools for food, clothing, and protection	• Processes of inventions and innovations • Specialization of labor • Evolution of techniques, measurement, and resources • Technological and scientific knowledge	• Evolutionary development of technology • Dramatic changes in society • History of technology • Early technological history • The Iron Age • The Middle Ages • The Renaissance • The Industrial Revolution • The Information Age
CHAPTER 5 DESIGN				
8 The Attributes of Design	• Everyone can design • Design is a creative process	• Definitions of design • Requirements of design	• Design leads to useful products and systems • There is no perfect design • Requirements	• The design process • Design problems are usually not clear • Designs need to be refined • Requirements
9 Engineering Design	• Engineering design process • Expressing design ideas to others	• Engineering design process • Creativity and considering all ideas • Models	• Iterative • Brainstorming • Modeling, testing, evaluating, and modifying	• Design principles • Influence of personal characteristics • Prototypes • Factors in engineering design
10 The Role of Troubleshooting, Research and Development, Invention and Innovation, and Experimentation in Problem Solving	• Asking questions and making observations • All products need to be maintained	• Troubleshooting • Invention and innovation • Experimentation	• Troubleshooting • Invention and innovation • Experimentation	• Research and development • Researching technological problems • Not all problems are technological or can be solved • Multidisciplinary approach
CHAPTER 6 ABILITIES FOR A TECHNOLOGICAL WORLD				
11 Apply the Design Process	• Solve problems through design • Build something • Investigate how things are made	• Collecting information • Visualize a solution • Test and evaluate solutions • Improve a design	• Apply design process • Identify criteria and constraints • Model a solution to a problem • Test and evaluate • Make a product or system	• Identify a design problem • Identify criteria and constraints • Refine the design • Evaluate the design • Develop a product or system using quality control • Reevaluate final solution(s)
12 Use and Maintain Technological Products and Systems	• Discover how things work • Use tools correctly and safely • Recognize and use everyday symbols	• Follow step-by-step instructions • Select and safely use tools • Use computers to access and organize information • Use common symbols	• Use information to see how things work • Safely use tools to diagnose, adjust, and repair • Use computers and calculators • Operate systems	• Document and communicate processes and procedures • Diagnose a malfunctioning system • Troubleshoot and maintain systems • Operate and maintain systems • Use computers to communicate

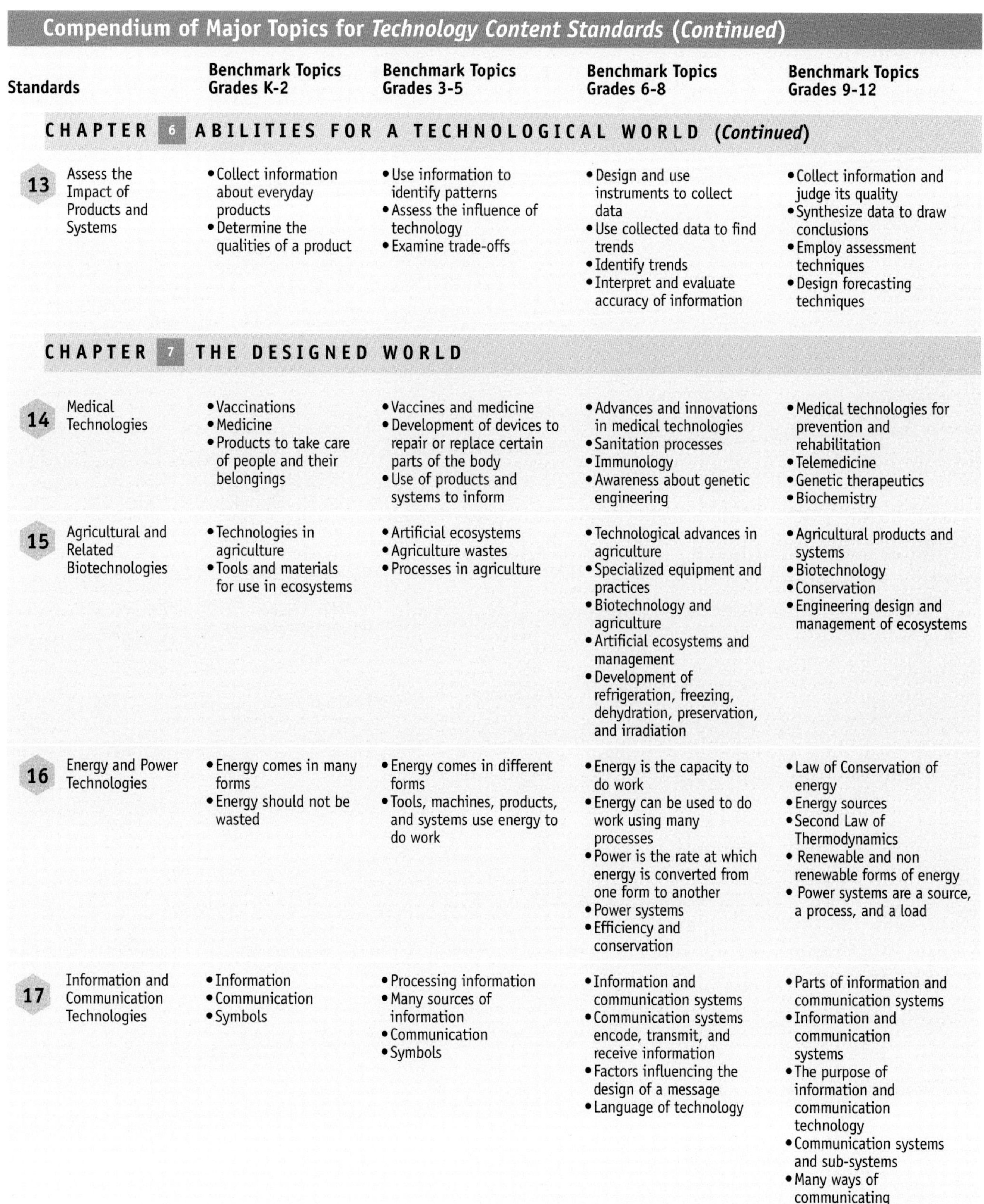

Compendium of Major Topics for *Technology Content Standards* (*Continued*)

Standards	Benchmark Topics Grades K-2	Benchmark Topics Grades 3-5	Benchmark Topics Grades 6-8	Benchmark Topics Grades 9-12
CHAPTER 6 ABILITIES FOR A TECHNOLOGICAL WORLD (*Continued*)				
13 Assess the Impact of Products and Systems	• Collect information about everyday products • Determine the qualities of a product	• Use information to identify patterns • Assess the influence of technology • Examine trade-offs	• Design and use instruments to collect data • Use collected data to find trends • Identify trends • Interpret and evaluate accuracy of information	• Collect information and judge its quality • Synthesize data to draw conclusions • Employ assessment techniques • Design forecasting techniques
CHAPTER 7 THE DESIGNED WORLD				
14 Medical Technologies	• Vaccinations • Medicine • Products to take care of people and their belongings	• Vaccines and medicine • Development of devices to repair or replace certain parts of the body • Use of products and systems to inform	• Advances and innovations in medical technologies • Sanitation processes • Immunology • Awareness about genetic engineering	• Medical technologies for prevention and rehabilitation • Telemedicine • Genetic therapeutics • Biochemistry
15 Agricultural and Related Biotechnologies	• Technologies in agriculture • Tools and materials for use in ecosystems	• Artificial ecosystems • Agriculture wastes • Processes in agriculture	• Technological advances in agriculture • Specialized equipment and practices • Biotechnology and agriculture • Artificial ecosystems and management • Development of refrigeration, freezing, dehydration, preservation, and irradiation	• Agricultural products and systems • Biotechnology • Conservation • Engineering design and management of ecosystems
16 Energy and Power Technologies	• Energy comes in many forms • Energy should not be wasted	• Energy comes in different forms • Tools, machines, products, and systems use energy to do work	• Energy is the capacity to do work • Energy can be used to do work using many processes • Power is the rate at which energy is converted from one form to another • Power systems • Efficiency and conservation	• Law of Conservation of energy • Energy sources • Second Law of Thermodynamics • Renewable and non renewable forms of energy • Power systems are a source, a process, and a load
17 Information and Communication Technologies	• Information • Communication • Symbols	• Processing information • Many sources of information • Communication • Symbols	• Information and communication systems • Communication systems encode, transmit, and receive information • Factors influencing the design of a message • Language of technology	• Parts of information and communication systems • Information and communication systems • The purpose of information and communication technology • Communication systems and sub-systems • Many ways of communicating • Communicating through symbols

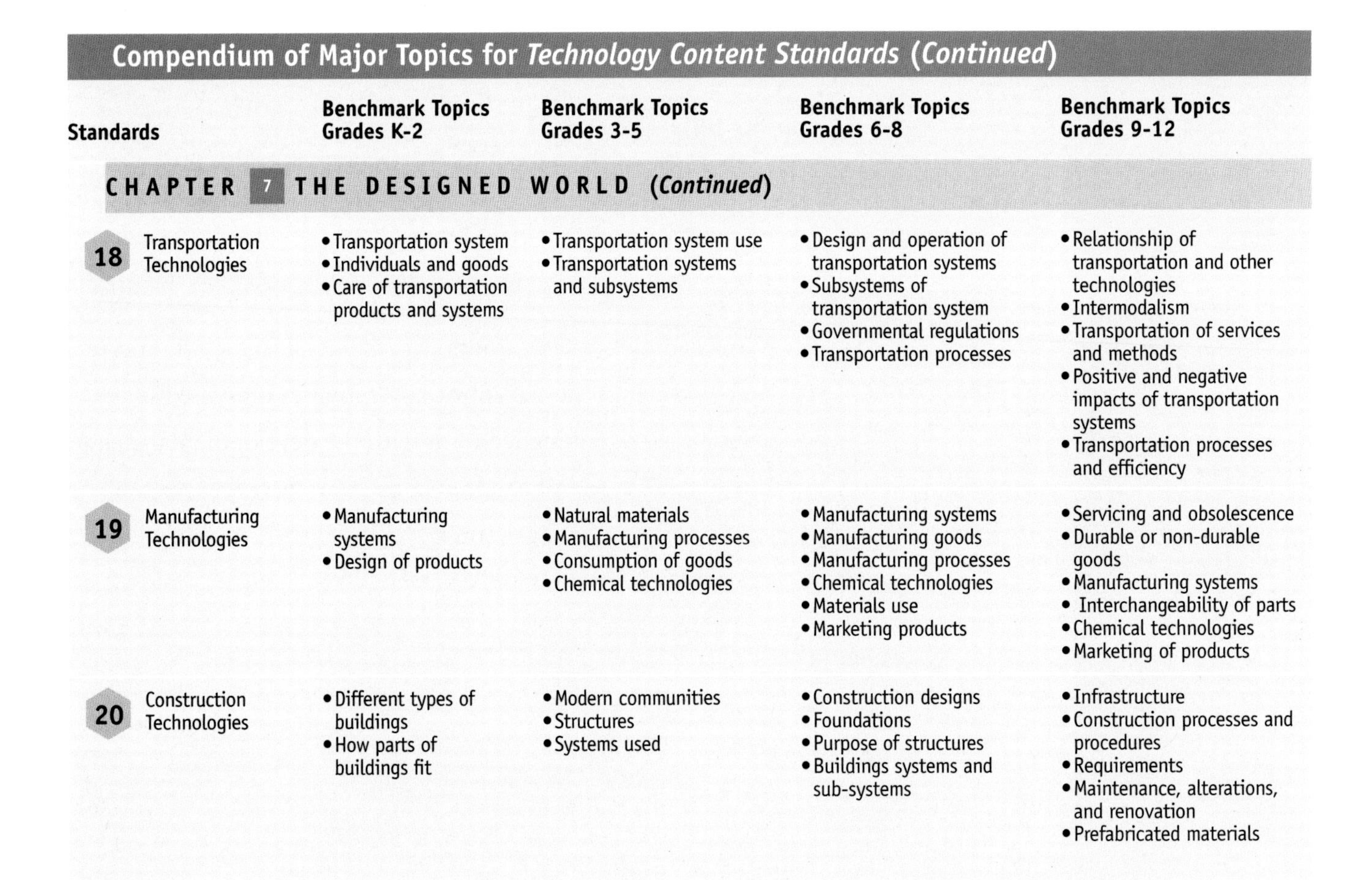

Compendium of Major Topics for *Technology Content Standards* (*Continued*)

Standards	Benchmark Topics Grades K-2	Benchmark Topics Grades 3-5	Benchmark Topics Grades 6-8	Benchmark Topics Grades 9-12
CHAPTER 7 THE DESIGNED WORLD (*Continued*)				
18 Transportation Technologies	• Transportation system • Individuals and goods • Care of transportation products and systems	• Transportation system use • Transportation systems and subsystems	• Design and operation of transportation systems • Subsystems of transportation system • Governmental regulations • Transportation processes	• Relationship of transportation and other technologies • Intermodalism • Transportation of services and methods • Positive and negative impacts of transportation systems • Transportation processes and efficiency
19 Manufacturing Technologies	• Manufacturing systems • Design of products	• Natural materials • Manufacturing processes • Consumption of goods • Chemical technologies	• Manufacturing systems • Manufacturing goods • Manufacturing processes • Chemical technologies • Materials use • Marketing products	• Servicing and obsolescence • Durable or non-durable goods • Manufacturing systems • Interchangeability of parts • Chemical technologies • Marketing of products
20 Construction Technologies	• Different types of buildings • How parts of buildings fit	• Modern communities • Structures • Systems used	• Construction designs • Foundations • Purpose of structures • Buildings systems and sub-systems	• Infrastructure • Construction processes and procedures • Requirements • Maintenance, alterations, and renovation • Prefabricated materials

D Articulated Curriculum Vignette from Grades K-12

One of the challenges of implementing *Technology Content Standards* is developing an articulated curriculum for grades K-12 that translates each of the standards into a planned curriculum with instructional activities suited for content being taught at each grade level. The following example illustrates how this could work in the laboratory-classroom.

This example revolves around the theme of transportation technology. Each of the transportation activities presented is age appropriate and designed to fit with the developmental characteristics and needs of children at the various grade levels. Also, each grade level builds upon the prior one, showing the importance of a continuous and articulated experience in technological studies from grades K-12.

GRADES
K-2

Students in grades K-2 exhibit a range of characteristics that influence the teaching and learning process in technological studies. These students need a wide variety of activities because they have short attention spans and tire easily, especially younger students. They are typically energetic and curious learners who enjoy cooperative learning activities that keep them active and allow them to use their rich imaginations. Because small muscles in the hands and fingers are not fully developed teachers must be cognizant of their students limited capabilities to do precise, manipulative tasks.

Technology activities in grades K-2 should address students' developmental characteristics, including their natural curiosity and inventive thinking skills. For example, students in grades K-1 should be given ample opportunities to explore and use wheels, axles, levers, gears, pulleys, and cams by playing with a variety of toys and construction kits that include these mechanisms. Older students can take apart, describe, and reassemble a simple toy vehicle or build a model of a conveyor system made with plastic building bricks.

By the end of grade 2, students should be able to design, plan, and make original vehicles using commercial construction kits and recyclable or consumable materials (e.g., boxes, straws, and craft sticks). Moreover, they should be able to use simple tools (e.g., hammers, scissors, and saws) safely and appropriately to accomplish their tasks.

When students explore mechanisms and design vehicles, they can sketch and describe these components and products to further enhance their understanding of the components' shapes, uses, and names.

In addition, younger students can organize mechanisms by characteristics including type, size, weight, and color to practice sorting tasks and to strengthen their skills in measurement and classification.

Throughout these grade levels, students can begin to use the terminology associated with *Technology Content Standards.* This vocabulary development can be acquired by involving students in activities that promote language development, such as orally presenting the projects and designs they have made, making a collage of transportation vehicles in various classifications (e.g., land, water, air, and space), and sketching and labeling drawings of devices they designed.

Teachers should clearly integrate technological studies with other areas of the curriculum throughout grades K-2. For example, connections to history and geography can be made by exploring the use of the inclined plane in the building of the Pyramids. Links to mathematics can be made by having students determine the rank of the vehicles they build (from slowest to fastest). Reading stories and engaging students in discussions about how their life could be different without cars and other powered vehicles could clearly support curriculum in language arts and social studies.

GRADES
3-5

Students in grades 3-5 are beginning to exhibit more individualism, but peer relationships are also important. These young learners have fully developed hand muscles and greatly improved hand-eye coordination that makes them more skillful at manipulative tasks requiring smaller and more numerous parts. Because their ability to stay focused on assignments is much improved, students are prepared to tackle design and problem-solving activities that require attention to greater detail for longer periods of time. As students mature, they become more capable of interpreting abstract concepts and making broad generalizations — essential traits for students being asked to evaluate designs and assess solutions to hypothetical, yet realistic, problems.

Activities in grades 3-5 should provide students with diverse opportunities to develop and enhance skills in designing, making, assessing, and presenting solutions to technological problems. Students can be challenged to use tools and materials for more ambitious tasks, such as creating vehicles that incorporate computer-controlled devices and use light, sound, or motion sensors. Using raw materials and simple hand tools, students can design, build, and test products that incorporate electricity, magnetism, and motors.

Problems at this level should build in complexity, and design constraints should become increasingly challenging. For example, students could build and use only one mechanism in their solutions at early levels, whereas students in grade 5 could incorporate as many as three mechanisms, each perhaps working in combination with the others. Students should more clearly articulate the assets and liabilities of their solutions and the positive and negative impacts of their designs.

In addition, fourth grade students could design and construct a model of a wastewater treatment system that moves and filters contaminated water (polluted with oil or containing sediments). Fifth grade students also could build and test hydraulic devices that simulate how the human body moves fluids.

All students at this level are capable of documenting their design and problem solving processes with conventional and computer-based sketching tools. Likewise, students should begin to use the World Wide Web (WWW) to display their designs (grades 3-4) and should document their problem-solving process through the use of a notebook or an electronic portfolio on the WWW (grade 5).

GRADES

6-8

Middle-level students are teetering between childhood and early adulthood. They are experiencing significant physical growth and change. Girls tend to mature earlier than boys, and they show markedly different interests from their male peers. These students are greatly influenced by their friends, and they typically reject adult guidance, making them more vulnerable to risky behaviors and more apt to present attitudinal difficulties. Likewise, these adolescents have an increased sense of self and a blossoming interest in members of the opposite gender. Placing middle-level students in teams allows them to have smaller communities for learning within the larger school and enables educational leaders to more effectively satisfy students' emotional and guidance needs as they make the transition to adulthood.

Students in grades 6-8 are ready to tackle more difficult technological problems that make them feel more mature and allow them to apply concepts and skills from other fields of study, such as mathematics and science. Small group activities may be particularly effective in building students' self-esteem by enabling boys and girls to have success with new and perplexing tasks. Constructing motors and generators from scratch or designing a game contraption that incorporates numerous simple machines and mechanisms to transport marbles or balls through a complicated maze may be well received at this level if students are given the opportunity to work in pairs or teams.

Middle-level students need to be challenged to refine skills learned in grades K-5 and to apply them to new problems and opportunities. Teachers should expect these students to take a more active role in formulating design problems and establishing constraints in order to ensure that activities reflect both male and female students' interests. Adolescents need to be given more decision-making roles, and they should be encouraged to have positive interactions with adults (e.g., parents, relatives, and other teachers) as they analyze their design options and assess the value and impacts of their solutions. In the process of compiling eletronic portfolios on the WWW or using another means of documenting their progress, these students will be able to clearly communicate how they transformed their ideas into practical solutions and how they appraised their solution's functional, aesthetic, social, and economic value. They also will use a variety of informational-technology tools, such as computer-aided design (CAD), multimedia software, database and spreadsheet applications, on-line search tools, and computer-control systems over networks, in order to accomplish these tasks.

GRADES

9-12

Students in grades 9-12 become increasingly independent while continuing to seek social acceptance. Their ability to think and visualize abstractly provides them with greater flexibility in problem solving. Physical maturation during these years results in greater size, dexterity, and strength as they mature into young men and women. They develop a clearer definition of their identity and their role in society, and they begin to formulate life ambitions and goals. Wage earning is often an immediate interest, with college and vocational decisions weighing heavily on their minds as they complete their high school "careers."

Technological studies at the high school level should take advantage of the particular interests students have during these years. Employment and career options, as well as consumer issues, are relevant topics for grades 9-12. Instructional activities should be challenging enough to hold their interest and encourage independent thinking as they pursue solutions to problems encountered along the way.

High school students are quite capable of developing sophisticated designs for research and development projects, so many activities at this level should have the look and feel of engineering projects. For example, they could develop a "smart" transportation system that employs computers and sensors. Their background research could require them to use on-line U.S. Patent Office searches and technical journals in their search for answers. They could also review the development of the interstate highway system in the United States. Students also might address social issues and conduct environmental impact studies associated with such systems.

In another scenario, students could work in "engineering teams" to design and build a space station simulation, giving full consideration to the variety of life-sustaining systems required in space. Or perhaps the simulation would be a virtual environment, existing in three dimensions on the WWW and accessible and controllable by students across the nation or on another continent.

A third approach might challenge students to design and build a solar-powered car for a state-wide competition. Along the way, students could explore social and environmental impacts, the working of solar cells, and the various subsystems at work within a car.

Summary

The study of technology could be addressed, in part, with a series of articulated activities from grades K-12 of increasing sophistication focusing on a particular theme. The transportation activities depicted in this example progress from imaginative play to exploratory design to sophisticated engineering. The activities recommended are developmentally appropriate, so that students will be challenged intellectually every step of the way. On this journey, students will experience a wide range of technologies in the context of real-world problems, thereby developing a very rich understanding of the technological world in which they live.

References For *Technology Content Standards*

Ackoff, R. (1978). *The art of problem solving.* New York: Wiley.

American Association for the Advancement of Science. (1989). *Science for all Americans.* New York: Oxford University Press.

American Association for the Advancement of Science, Project 2061. (1989). *Phase I physical and information sciences and engineering panel report.* Washington, DC: Author.

American Association for the Advancement of Science. (1993). *Project 2061: Benchmarks for science literacy.* New York: Oxford University Press.

Armstrong, T. (1994). *Multiple intelligences in the classroom.* Alexandria, VA: Association for Supervision and Curriculum Development.

Ausubel, D. P. (1967). Learning theory and classroom practice. *The Ontario Institute for Studies in Education* (Bulletin No. 1). Toronto, Canada: The Ontario Institute for Studies in Education.

Ausubel, D. P., Sullivan, E. V., & Ives, S. W. (1980). *Theory and problems of child development.* New York: Grune & Stratton.

Banks, F. (Ed.). (1994). *Teaching technology.* London: Routledge.

Barnes, J. L. (1987). *An international study of curriculuar organizers for the study of technology.* Unpublished doctoral dissertation, Virginia Polytechnic Institute and State University, Blacksburg.

Basalla, G. (1988). *The evolution of technology.* New York: Cambridge University Press.

Boser, R. A. (1993). The development of problem solving capabilities in pre-service technology teacher education. *Journal of Technology Education, 4*(2), 12-29.

Brauer, W. M., & Bensen, M. J. (1997). *Tomorrow's technology: 'Know-how' for manufacturing and technology leaders of the 21st century* (Technical. Rep.). Bemidji, MN: Bemidji State University, Minnesota Technology, Inc.

Bronzino, J. D. (Ed.). (1995). *The biomedical engineering handbook.* Boca Raton, FL: CRC Press.

Bruer, J. (1993). *Schools for thought: A science of learning in the classroom.* London: The MIT Press.

Bruner, J. S. (1967). *Toward a theory of instruction.* Cambridge, MA: Belknap Press of Harvard University Press.

Bruner, J., & Haste, H. (1987). *Making sense: The child's construction of the world.* New York: Methuen.

Brusic, S. A. (1991). *Determining the effects on fifth grade students' achievement and curiosity when a technology education activity is integrated with a unit in science.* Unpublished doctoral dissertation, Virginia Polytechnic Institute and State University, Blacksburg.

Brusic, S. A., & Barnes, J. L. (1992). *Kids & technology: Mission 21, launching technology across the curriculum, level 3.* New York: Delmar.

Bugliarello, G. (1990). *The intelligent layman's guide to technology.* New York: Polytechnic Press.

Burrus, D. (1993). *Technotrends: How to use technology to go beyond your competition.* New York: Harper Business.

Childress, V. W. (1994). *The effects of technology education, science, and mathematics integration upon eighth grader's technological problem-solving ability.* Unpublished doctoral dissertation, Virginia Polytechnic Institute and State University, Blacksburg.

Childress, V. W. (1996). Does integrating technology, science, and mathematics improve technological problem solving? A quasi-experiment. *Journal of Technology Education, 8* (1), 16-26.

Coenen-Van Den Bergh, R. (Ed.). Report PATT-conference 1987 Vol. 2 Contributions. *Pupil's Attitude Towards Technology.* Eindhoven, The Netherlands: Bariet, Ruinen.

Committee on Life Sciences and Health of the Federal Coordinating Council for Science, Engineering, and Technology. (1992). *Biotechnology for the 21st century: Realizing the promise.* Washington, DC: U.S. Government Printing Office.

Committee on Life Sciences and Health of the Federal Coordinating Council for Science, Engineering, and Technology. (1993). *Biotechnology for the 21st century: Realizing the promise.* Washington, DC: U.S. Government Printing Office.

Connecticut State Department of Education Division of Teaching and Learning. (1998). *Technology education curriculum framework.* Connecticut: Author.

Croft, V. E. (1990). *A national study to determine the characteristics of technological literacy for high school graduates.* Unpublished doctoral dissertation, Virginia Polytechnic Institute and State University, Blacksburg.

Cropley, A. (1992). *More ways than one: Fostering creativity.* Norwood, NJ: Ablex.

Cross, N. (1994). *Engineering design methods: Strategies for product design* (3rd ed.). Chichester, England: John Wiley & Sons.

Custer, R. L. (1991). *Technology: A qualitative concept analysis from the perspectives of engineering, philosophy, natural science, and technology education.* Unpublished doctoral dissertation, University of Missouri - Columbia, Columbia, Missouri.

Custer, R. L., & Weins, A. E. (Eds.) (1996). *Technology and the quality of life:* 45th yearbook, Council on Technology Teacher Education. New York: Glencoe/McGraw-Hill.

Damon. W. (Ed.). (1991). *Child development today and tomorrow.* Oxford: Jossey-Bass.

Dasgupta, S. (1996). *Technology and creativity.* New York: Oxford University Press.

de Klerk Wolters, F., Mottier, I., Raat, J. H., & de Vries, M. (Eds.). (1989). Teacher education for school technology. *Pupil's Attitude Towards Technology.* (Report PATT-4 conference). Eindhoven, The Netherlands.

DeLuca, V.W. (1991). Implementing technology education problem-solving activities. *Journal of Technology Education, 2*(2), 5-15.

Department for Education and Welsh Office Education Department. (1995). *Design and technology in the national curriculum.* London: HMSO.

DeVore, P. W. (1980). *Technology: An introduction.* Worcester, MA: Davis.

De Vries, M.J., & Grant, D.P. (Eds.). (1993). *Design methodology and relationships with science.* The Netherlands: Kluwer Academic Publishers.

De Vries, R. & Zan, B. (1994). *Moral classrooms, moral children: Creating a constructivist atmosphere in early education.* New York: Teachers College Press.

Dillon, R. F., & Sternberg, R. J. (Eds). (1986). *Cognition and instruction.* New York: Academic Press.

Drake, S. M. (1993). *Planning integrated curriculum: The call to adventure.* Alexandria, VA: Association for Supervision and Curriculum Development.

Dugger, W. E., Jr., Bame, A. E., Pinder, C. A., & Miller, D. C. (1985). *Standards for technology education programs.* Reston, VA: International Technology Education Association.

Dugger, W. E., Jr. (1988). Technology — The discipline. *The Technology Teacher, 48*(1), 3-6.

Dugger, W. E., Jr., & Yung, J. E. (1995). Technology education today. *Fastback, 380,* Bloomington, ID: Phi Delta Kappa Educational Foundation.

Dunlop, D. D., Croft, V. E., & Brusic, S. A. (1992). *Kids & Technology: Mission 21, launching technology across the curriculum, Level 2.* Albany, NY: Delmar Publishers Inc.

Dunn, S., & Larson, R. (1990). *Design technology: Children's engineering.* New York: Falmer.

Dyrenfurth, M. J., & Kozak, M. R. (1991). *Technological literacy:* 40th yearbook, Council on Technology Teacher Education. Peoria, IL: Glencoe Division, Macmillan/McGraw-Hill.

Ferguson, E. S. (1962). On the origin and development of American mechanical 'know-how.' *Journal of the Central Mississippi Valley American Studies, 3,* 3-16.

Ferguson, E. S. (1977). The mind's eye: Nonverbal thought in technology. *Science, 197,* 827-836.

Fleming, R. (1986a). Adolescent reasoning in socio-scientific issues, Part I: Social cognition. *Journal of Research in Science Teaching, 23* (8), 677-687.

Fleming, R. (1986b). Adolescent reasoning in socio-scientific issues, Part II: Nonsocial cognition. *Journal of Research in Science Teaching, 23* (8), 689-698.

Fleming, R. W. (1987). High-school graduate's beliefs about science-technology-society. Part II: The interaction among science, technology and society. *Science Education 71* (2), 163-186.

Florman, S. (1987). *The civilized engineer.* New York: St. Martin's Press.

Florman, S. C. (1994). *The existential pleasures of engineering* (2nd ed.). New York: St. Martin's Griffin.

Foster, W. T. (1992). Technology education research: Looking to the future. *The Technology Teacher,* pp. 33-34.

French, M. (1994). *Invention and evolution: Design in nature and engineering* (2nd ed.). New York: Cambridge University Press.

Gagel, C. W. (1997). Literacy and technology: Reflections and insights for technological literacy. *Journal of Industrial Teacher Education 34* (3), 6-34.

Gagne, R. M., & Glaser, R. (1987). Foundations in learning research. In Robert M. Gagne (Ed.). *Instructional technology foundations* (pp. 49-84). Hillsdale, NJ: Lawrence Erlbaum Associaties.

Gagne, R. M., & Driscoll, M. P. (1988). *Essentials of learning for instruction* (2nd. ed.). Englewood Cliffs, NJ: Prentice Hall.

Gardner, H. (1993). *Frames of mind.* New York: BasicBooks.

Gardner, H. (1993). *Creating minds: An anatomy of creativity seen through the lives of Freud, Einstein, Picasso, Stravinsky, Eliot, Graham, and Gandhi.* New York: BasicBooks.

Gates, B. (1995). *The road ahead.* New York: Viking Penguin.

Geography Standards Education Project. (1994). *Geography for life: National geography standards.* Washington, DC: National Geographic Research & Exploration.

Gibson, J. E. (1968). *Introduction to engineering design.* New York: Holt, Rinehart and Winston, Inc.

Gilberti, A. F. (1989). Technological literacy as a curriculum movement (Doctoral dissertation, University of Maryland-College Park, 1989). *Dissertation Abstracts International, 50-07A,* 1967-2200.

Goetsch, D.L. & Nelson, J.A. (1987). *Technology and you.* Albany, NY: Delmar Publishers Inc.

Government of Newfoundland and Labrador Department of Education and Training Division of Program Development. (1998). *Technology education outcomes.* Manuscript in preparation.

Gradwell, J., Welch, M., & Martin, E. (1991). *Technology: Shaping our world*. South Holland, IL: Goodheart-Willcox.

Hacker, M. & Barden, R.A. (1993). *Living with technology.* New York: Delmar Publishers Inc.

Hall, A. R. (1983). *The revolution in science, 1500-1800.* London: Longman.

Healy, J. M. (1990). *Endangered minds: Why our children don't think*. New York: Simon & Schuster.

Hellweg, P., (1997). *The American heritage children's thesaurus*. Boston, MA: Houghton Mifflin Company.

Henderson, J., & Knutton, S. (1990). *Biotechnology in schools: A handbook for teachers.* Philadelphia: Open University Press.

Hendricks, R.W. & Sterry, L.F. (1996*). Communication today.* Menomonie, WI: T & E Publications.

Herman, J. L., Aschbacher, P. R., & Winters, L. (1992). *A practical guide to alternative assessment*. Alexandria, VA: Association for Supervision and Curriculum Development.

Hickman, L. A. (1992). *John Dewey's pragmatic technology*. Bloomington, IN: Indiana University Press.

Hiebert, J., Carpenter, T. P., Fennema, E., Fuson, K., Human, P., Murray, H., Olivier, A., & Wearne, D. (1996). Problem solving as a basis for reform in curriculum and instruction: The case of mathematics. *Educational Researcher*, 12-21.

Hill, P. H. (1970). *The science of engineering design*. New York: Holt, Rinehart and Winston, Inc.

Hofer, B., & Pintrich, P. R. (1997, Spring). The development of epistemological theories: Beliefs about knowledge and knowing and their relation to learning. *Review of Educational Research, 67,* 88-140.

Householder, D.L. (Ed). (1972). *Industrial arts for the early adolescent*: 21st yearbook. American Council on Industrial Arts Teacher Education. Bloomington, IL: McKnight & McKnight Publishing Company.

Industrial Arts Curriculum Project. (1970). *The World of Construction.* Bloomington, IL: McKnight & McKnight Publishing Company.

Industrial Arts Curriculum Project. (1971). *The World of Manufacturing.* Bloomington, IL: McKnight & McKnight Publishing Company.

Inhelder, B., & Piaget, J. (1958). *The growth of logical thinking: From childhood to adolescence.* France: Presses Universitaires de France.

Israel, E. N., & Wright, R. T. (Eds.). (1987). *Conducting technical research:* 36th yearbook, Council on Technology Teacher Education. Mission Hills, CA: Glencoe.

Jacobs, H.H. (1997, March/April). Designing with rigor: Crafting interdisciplinary high school curricula. *The High School Magazine, 4*, 32-39.

Jenkins, Edgar W. (1997, Winter/Spring). Technological literacy: Concepts and constructs. *The Journal of Technology Studies, XXIII,* 2-5.

Jewkes, J., Sawers, D., & Stillerman, R. (1958). *The sources of invention.* New York: Macmillan.

Johnson, C., & Wellman, H. (1982). Children's developing conceptions of the mind and brain. *Child Development, 53* (1), 222-234.

Johnson, D. W., Johnson, R. T., & Holubec, E. J. (1994). *The new circles of learning: Cooperation in the classroom and school.* Alexandria, VA: Association for Supervision and Curriculum Development.

Johnson, J. R. (1993). *Technology: Report of the project 2061 phase I technology panel.* Washington, DC: American Association for the Advancement of Science.

Katz, L. G. (1996, June). Child development knowledge and teacher preparation: Confronting assumptions. *Early Childhood Research Quarterly, 11,* 135-146.

Keller, C. M., & Keller, J. D. (1996). *Cognition and tool use: The blacksmith at work.* New York: Cambridge University Press.

Knorr-Cetina, K. (1981). *The manufacture of knowledge: An essay on the constructivist and contextual nature of science.* New York: Pergamon Press.

Kuhn, T. S. (1970). *The structure of scientific revolutions* (2nd ed.). Chicago: The University of Chicago Press.

LaPorte, J. E., & Sanders, M. E. (1995). *Technology, science, mathematics connection activities.* Peoria, IL: Glencoe/McGraw-Hill.

LaPorte, J. E., & Sanders, M. E. (1995). Technology, science, mathematics integration. In G. Martin (Ed.), *Foundations of technology education:* 44th yearbook, Council on Technology Teacher Education. Peoria, IL: Glencoe/McGraw-Hill.

Layton, E. T., Jr. (1974). Technology as knowledge. *Technology and Culture, 15*, 31-41.

Lee, L., Wang, W., Chen, N., Lin, F., Tsai, S., & Peng, Y. (1997). *An introduction to technology education in the Republic of China on Taiwan.* Department of Industrial Technology Education, National Taiwan Normal University.

Lee, L. (1998). *A comparison of technology education programs in eight Asia-Pacific countries.* Paper presented at the Austrailian Council of Education Through Technology (ACET), national conference, Melbourne, Victoria, Australia.

Liedtke, Jane A. (Ed.). (1990). *Communication in technology education:* 39th yearbook, Council on Technology Teacher Education. Mission Hills, CA: Glencoe/McGraw-Hill.

Lindberg, D. C. (1992). *The beginnings of western science.* Chicago: The University of Chicago Press.

Lisensky, R. P., Pfnister, A. O., & Sweet, S. D. (1985). *The new liberal learning: Technology and the liberal arts.* Washington, D. C.: The Council of Independent Colleges.

Loepp, F., Fisher, R., Meier, S., Hovde, R., Shea, G., Shook, S., & Williamson, V. (1998). *Wellness (student edition, teacher edition, journal sheets),* Peoria: Glencoe McGraw-Hill.

Loepp, F., Fisher, R., Meier, S., Hovde, R., Shea, G., Shook, S., & Williamson, V. (1998). *Food production* (student edition, teacher edition, journal sheets), Peoria: Glencoe McGraw-Hill.

Macaulay, D. (1998). *The new way things work.* Boston: Houghton Mifflin.

MacQuitty, J. J. (1997, April). The real implications of Dolly. *Nature Biotechnology,* 294.

Maley, D. (1985). *Math/science/technology.* Reston, VA: International Technology Education Association.

Markert, L. R. (1993). *Contemporary technology: Innovations, issues, and perspectives.* South Holland, IL: Goodheart-Willcox.

Martinez, M. E. (1998). What is problem solving? *Phi Delta Kappan, 79* (8), Bloomington, IN: Phi Delta Kappa International.

Marzano, R., & Kendall, J. (1996). *A comprehensive guide to designing standards-based districts, schools and classrooms.* Aurora, CO: Mid-Continent Regional Educational Laboratory.

Michigan Department of Education. (1998). *Technology education curriculum guide.* Lansing, MI: Author.

Mid-Continent Regional Educational Laboratory Standards Database (1998). Technology standards [On-line]. Available: *http://www.mcrel.org/standards-benchmarks/docs/chapter18.html*

Middendorf, W. H. (1969). *Engineering design.* Boston: Allyn and Bacon, Inc.

Middendorf, W.H., & Engelmann, R.H. (1998). *Design of devices and systems* (3rd ed.). New York: Marcel Dekker.

Middleton, H. E. (1990). Technology and Science: Making the links. In M. Dupe (Ed.). *Making the links: Technology and science, industry, and education.* (pp. 5-30). Canberra: AGPS.

Middleton, H. E., (1997). Working harder or working smarter: Technology education in the twenty-first century. *Report of the 1997 International Conference of Technology Education in the Asia-Pacific Region.* Taipei: National Taiwan Normal Univsersity.

Middleton, H. E. (Ed.). (1992-1999). Design in education. *Journal of the Design in Education Council,* Australia.

Miles, Jr., R.F. (Ed.). (1973). *Systems concepts: Lectures on contemporary approaches to systems.* New York: John Wiley & Sons.

Mitcham, C., (1994). *Thinking through technology: The path between engineering and philosophy.* Chicago: The University of Chicago Press.

Mitcham, C. (Ed.) (1995). *Technology's school: The challenge to philosophy.* Greenwich, CT: JAI Press.

Miyakawa, H., & Nakashima, Y. (1996). *Study on the fostering of creativity in technology education: Comparison of average scores between male and female students.* Kariya-City, Japan: Department of Technology Education, Aichi University of Education.

Morris, C. (Ed.). (1996). *Academic Press Dictionary of Science and Technology.* New York: Harcourt-Brace Jovanovich.

Mottier, I., Raat, J. H., & de Vries, M. (Eds.). (1991). Technology education and industry. *Pupil's Attitude Towards Technology.* (Report PATT-5 conference, Vol. 1). Eindhoven, The Netherlands: PATT-Foundation.

Mottier, I., Raat, J. H., & de Vries, M. (Eds.). (1993). Technology education and the environment: Improving our environment through technology education. *Pupil's Attitude Towards Technology.* (Report Proceedings PATT-6 conference). Eindhoven, The Netherlands: PATT-Foundation.

National Academy of Engineering, (1989). *Technology and environment.* Washington, DC: National Academy Press.

National Council of Teachers of Mathematics. (1989). *Curriculum and evaluation standards for school mathematics.* Reston, VA: Author.

National Research Council. (1996). *National science education standards.* Washington, DC: National Academy Press.

National Research Council (1999). *How people learn: Brain, mind, experience, and school.* Washington, DC: National Academy Press, p57.

New York State Education Department, (1996). *Learning standards for mathematics, science, and technology.* New York: Author.

Novak, J. D. (1998). *Learning, creating, and using knowledge: Concept maps™ as facilitative tools in schools and corporations.* Mahwah, NJ: Lawrence Erlbaum Associates.

Novotny, J. A. (1995). *A review of problem-solving approaches related to the study of technology and means to assess student problem-solving ability.* Unpublished manuscript.

Office of Technology Assessment of the Congress of the United States (1988). *U.S. investment in biotechnology – Special report.* Boulder, CO: Westview Press.

Office of Technology Assessment of the Congress of the United States (1991). *U.S. investment in biotechnology – Special report.* Boulder, CO: Westview Press.

Pacey, A. (1990). *Technology in world civilization: A thousand-year history.* Cambridge, MA: The MIT Press.

Parayil, G. (1991). Technological knowledge and technological change. *Technology in Society, 13,* 289-304.

Pennsylvania Department of Education. (1998). *Proposed academic standards for science and technology.* Manuscript in preparation.

Petroski, H. (1990). *The pencil: A history of design and circumstance.* New York: Alfred A. Knopf.

Petroski, H. (1993). *The evolution of useful things.* New York: Alfred A. Knopf.

Piaget, J. (1952). *The origins of intelligence in children* (M. Cook, Trans.). New York: International Universities Press.

Polya, G. (1962-1965). *Mathematical discovery: On understanding, learning, and teaching problem solving* (Vol. 1-2). New York: John Wiley & Sons.

Postman, N. (1995). *The end of education: Redefining the value of school.* New York: Alfred A. Knopf.

Pretzer, W. S. (1996, June 24). *Why learn technology: The quest for truth, beauty, and love.* Paper presented at the meeting of the Technical Foundation of America, Lahaina, Maui, Hawaii.

Pytlik, E. C. (1997, March). Conducting qualitative research in the classroom. *The Technology Teacher,* 20-21.

Pytlik, E. C., Lauda, D. P., & Johnson, D. L. (1978). *Technology, change and society.* Worcester, MA: Davis.

Raat, J. H., & de Vries, M. (Eds.). (1986). What do girls and boys think of technology? *Pupil's Attitude Towards Technology.* (Report of the PATT-workshop). Eindhoven, The Netherlands: Bariet, Ruinen.

Raat, J. H., Coenen-Van Den Berh, R., de Klerk Wolters, F., & de Vries, M. (Eds.). (1988). Basic principles of school technology. *Pupil's Attitude Towards Technology.* (Report PATT-3 conference, Vol. 2). Eindhoven, The Netherlands: Bariet, Ruinen.

Raat, J., de Vries, M., & Mottier, I. (Eds.). (1995). Teaching technology for entrepreneurship and employment. *Pupil's Attitude Towards Technology.* (Report Proceedings PATT-7 conference). Western Cape, South Africa: Nasou-Via Afrika.

Raizen, S., Sellwood, P., Todd, R., & Vickers, M. (1995). *Technology education in the classroom: Understanding the designed world.* San Francisco: Jossey-Bass.

Ritz, J.M., Hadley, W.F., & Bonebrake, J. (1990). *Exploring production systems: Processing, construction, manufacturing.* Worcester, MA: Davis Publications.

Robinson, A., & Stern, S. (1997). *Corporate creativity: How innovation and improvement actually happen.* San Francisco: Berrett-Koehler.

Rossner, A. G. (1990). *Validation of a bio-related technology taxonomy.* Unpublished master's thesis, Bowling Green State University, Bowling Green, Ohio.

Rutherford, J. & Ahlgren, A. (1990). *Science for all Americans.* New York: Oxford University Press.

Sahal, D. (1981). *Patterns of technological innovation.* Reading, MA: Addison-Wesley.

Sanders, M. E. (1991). *Communication technology: Today and tomorrow.* Mission Hills, CA: Glencoe/McGraw Hills Publishing Co.

Sanders, M. E. (1993). Science and technology: A new alliance. *Science Scope, 16* (6), 56-60.

Sanders, M. E. (1994). Technological problem-solving activities as a means of instruction. *School Science and Mathematics, 94* (1).

Santrock, J. W. (1996). *Child development.* (7th ed.). Dubuque, IA: Brown & Benchmark.

Savage, E., & Sterry, L. (1990). *A conceptual framework for technology education.* Reston, VA: International Technology Education Association.

Schauble, L., Klopfer, L. E., & Raghavan, K. (1991). Students' transition from an engineering model to a science model of experimentation. *Journal of Research in Science Teaching, 28,* 859-882.

Schmitt, M.L. & Pelley, A.L. (1966). *Industrial arts education: A survey of programs, teachers, students, and curriculum.* Washington, DC: U.S. Government Printing Office.

Scientific American. (1971). *Energy and power.* San Francisco: W.H. Freeman and Company.

Senge, P. (1990). *The fifth discipline: The art and practice of the learning organization.* New York: Doubleday Currency.

Seymour, R. D., & Shackelford, R. L. (Eds.). (1993). *Manufacturing in technology education:* 42nd yearbook, Council on Technology Teacher Education. Mission Hills, CA: Glencoe/McGraw-Hill.

Shepard, S.B. (Ed.). (1999, Summer). 100 years of innovation: A photographic journey. *Business Week.*

Shulman, L. S. (February, 1986). Those who understand: Knowledge growth in teaching. *Educational Researcher, 15,* 4-14.

Skolimowski, H. (Summer, 1966). The structure of thinking in technology. *Technology and Culture, VII (3),* 371-383.

Smith, H.B. (1985). *Exploring energy sources/applications/alternatives.* South Holland, IL: The Goodheart-Willcox Company, Inc.

Snyder, J. F., & Hales, J. A. (Eds.). (1981). *Jackson's Mill industrial arts curriculum theory.* Reston, VA: International Technology Education Association.

Tishman, S., & Perkins, D. (1997). The language of thinking. *Phi Delta Kappan, 78,* 368-374.

Technology in the New Zealand curriculum (1995). Wellington, New Zealand: Learning Media Limited.

Technology Education Advisory Council. (1988). *Technology: A national imperative.* Reston, VA: International Technology Education Association.

Technology for All Americans Project. (1996). *A rationale and structure for the study of technology.* Reston, VA: International Technology Education Association.

The Design Council, (1995). *Definitions of design.* London: The Design Council.

The International Society for Technology in Education. (1998). *National educational technology standards for students.* [On-line]. Available: http://cnets.iste.org/

The National Council for the Social Studies. (1994). *Expectations of excellence: Curriculum standards for social studies.* Washington, D.C.: Author.

Todd, R.D., McCrory, D.L., & Todd, K.I. (1985). *Understanding and using technology.* Worcester, MA: Davis Publications, Inc.

Torp, L., & Sage, S. (1998). *Problems as possibilities: Problem-based learning for K-12 education.* Alexandria, VA: Association for Supervision and Curriculum Development.

Torrance, E. P. (1992, January/February). A national climate for creativity and invention. *Gifted Child Today,* 10-14.

U. S. Department of Labor, The Secretary's Commission on Achieving Necessary Skills. (1991). *What work requires of schools, A SCANS report for America 2000.* (DOC Publication No. PB92-146711). Springfield, VA: National Technical Information Service.

Vincenti, W. G. (1990). *What engineers know and how they know it: Analytical studies from aeronautical history.* Baltimore, MD: The Johns Hopkins University Press.

Vygotsky, L. (1986). *Thought and language.* London: The MIT Press.

Waetjen, W. B. (1989). *Technological problem solving: A proposal.* Reston, VA: International Technology Edcuation Association.

Waetjen, W. B. (1993). Technology literacy reconsidered [16 paragraphs]. *Journal of Technology Education* [On-line serial], 4(2). Available On-line: http://scholar.lib.vt.edu/ejournals/JTE/jte-4n2/waetjen.jte-v4n2.html

Waetjen, W. B. (1995). Language: Technology in another dimension. *Journal of Technology Studies, 21,* 2.

Walker, J.R. (1976). *Exploring power technology basic fundamentals.* South Holland, IL: The Goodheart-Willcox Company, Inc.

Warner, W. (1965). *A curriculum to reflect technology.* American Industrial Arts Association Feature Presentation, April 25th, 1947. Columbus: Epsilon Pi Tau, Inc.

Watson, B., & Konicek, R. (May, 1990). Teaching for conceptual change: Confronting children's experience. *Phi Delta Kappan, 71,* 680-685.

Weber, R. (1992). *Forks, phonographs, and hot air balloons: A field guide to inventive thinking.* New York: Oxford University Press.

Weber, R., & Perkins, D. (Eds.). (1992). *Inventive minds: Creativity in technology.* New York: Oxford University Press.

Welch, M. W. (1996). *The strategies used by ten grade 7 students, working in single-sex dyads, to solve a technological problem.* Unpublished doctoral dissertation, McGill University, Montreal, Canada.

Welch, M., & Lim, H. S. (1998). *The effect of problem type on the strategies used by novice designers.* Manuscript submitted for publication.

Welch, M. (1998). *Students' use of three-dimensional modelling while designing and making a solution to a technological problem.* Manuscript submitted for publication.

Wells, J.G. (1994). Establishing a taxonometric structure for the study of biotechnology in secondary school technology education. *Journal of Technology Education, 6* (1), 58-75.

Wells, J. G. (1992). Establishment of a taxonometric structure for the study of biotechnology as a secondary school component of technology education. *Dissertation Abstracts International, 53-07A,* 2275-2522.

Wells, J.G. & Brusic, S.A. (1992). *Mission 21: Launching technology across the curriculum.* New York: Delmar Publishers, Inc., 128-131.

Wells, J. G., & Brusic, S. (1993). *Kids & technology: Mission 21, launching technology across the curriculum, Level 1.* New York: Delmar.

Wells, J. G. (1994, Fall). Establishing a taxonometric structure for the study of biotechnology in secondary school technology education. *Journal of Technology Education, 6* (1), 58-75.

Wescott, J. W., & Henak, R. M. (Eds.). (1994). *Construction in technology education:* 43rd yearbook, Council on Technology Teacher Education. Mission Hills, CA: Glencoe/McGraw-Hill.

Wickelgren, W.A. (1974). *How to solve problems: Elements of a theory of problems and problem solving.* San Francisco: W.H. Freeman and Company.

Wiggins, G., & McTighe, J. (1998). *Understanding by design.* Alexandria, VA: Association for Supervision and Curriculum Development.

Williams, P., & Jinks, D. (1985). *Design and technology 5-12.* Philadelphia: The Falmer Press.

Woodson, T. T. (1966). *Introduction to engineering design.* Los Angeles: McGraw-Hill, Inc.

Wright, J. R., & Komacek, S. A. (Eds.). (1992). *Transportation in technology education:* 41st yearbook, Council on Technology Teacher Education. Mission Hills, CA: Glencoe/McGraw-Hill.

Wright, R. T., & Smith, H. B. (1998). *Understanding technology.* Tinley Park, IL: Goodheart-Willcox.

Wulf, W. A. (1998, June 28). Education for an age of technology. *Op-Ed Service, National Academy of Engineering, National Academy of Sciences, and the Institute of Medicine.*

Zuga, K. F. (1994). *Implementing technology education: A review and synthesis of the research literature* (Information Series No. 356). Washington, D. C.: Office of Educaitonal Research and Improvement. (ERIC Document Reproduction Service No. ED 372 305).

Acknowledgements

The following lists have been compiled as carefully as possible from our records. We apologize to anyone whom we have omitted or whose name, title, or affiliation is incorrect. Inclusion on these lists does not imply endorsement of this document.

Technology for All Americans Project Staff

William E. Dugger, Jr., DTE, Director

Pam B. Newberry, Researcher & Contributing Writer

Melissa Smith, Editor

Stephanie Overton, Publications Coordinator

Constance Moehring, Volunteer Librarian & Researcher

Crystal Nichols, Editorial Assistant & Data Coordinator

International Technology Education Association Staff

Kendall Starkweather, DTE, Executive Director

Thomas Hughes, Jr., Past Director of Foundation Development

Brigitte Valesey, DTE, Director of Professional Development

Katie de la Paz, Advertising/Marketing Coordinator

Michele Judd, Meeting Planning Coordinator

Lee Anne Pirrello, Exhibits Coordinator

Phyllis Wittmann, Accounting Coordinator

Lari Price, Member Services Coordinator

Catherine James, Administrative/Website Coordinator

Moira Wickes, Database Coordinator/Registrar

Kathie Cluff, Publications Services Coordinator

Barbara Mongold, Public Services Coordinator

International Technology Education Association Board of Directors

Anthony Gilberti, DTE, President, Indiana State University, Indiana

Barry Burke, DTE, President-Elect, Montgomery County Public Schools, Maryland

Ronald Yuill, DTE, Past-President, Tecumseh Middle School, Indiana

Kendall Starkweather, DTE, Executive Director, ITEA, Virginia

William Havice, DTE, TECA Director, Clemson University, South Carolina

James Kirkwood, DTE, TECC Director, Ball State University, Indiana

Everett Israel, DTE, CTTE Director, Eastern Michigan University, Michigan

Harold Holley, ITEA-CS Director, Oklahoma Department of Vocational and Technical Education, Oklahoma

Thomas Bell, Region 1 Director, Millersville University, Pennsylvania

Gary Wynn, DTE, Region 2 Director, Greenfield-Central High, Indiana

Duane Rogers, Region 3 Director, Eastern Hills High School, Texas

Dean Christensen, Region 4 Director, Davis School District, Utah

Standards Team

Grades K-2 and 3-5

Jane Wheeler, Monte Vista Elementary, California, Leader

Michael Wright, DTE, University of Missouri-Columbia, Missouri, Recorder

Clare Benson, University of Central England, United Kingdom

Kristin Callender, Deane Elementary, Colorado

Linda S. Hallenbeck, East Woods School, Ohio

Jane Hill, Brazosport Independent School District, Texas

Stephan Knobloch, Crossfield Elementary, Virginia

Connie Larson, John Wetton Elementary, Oregon

Kathy Thornton, University of Virginia, Virginia

Grades 6-8

Franzie Loepp, DTE, Illinois State University, Illinois, Leader

Brigitte Valesey, DTE, Center to Advance the Teaching of Technology & Science (CATTS), Virginia, Recorder

William Ball, DTE, Clague Middle, Michigan

Barry Burke, DTE, Montgomery County Public Schools, Maryland

Denise Denton, University of Washington, Washington

Michael Hacker, The State University of New York at Stony Brook, New York

Chip Miller, Century High, Oregon

Tonia Schofield, Sylvan Middle, Georgia

Leon Trilling, Massachusetts Institute of Technology, Massachusetts

Grades 9-12

Rodney Custer, DTE, Illinois State University, Illinois, Leader

Anthony Gilberti, Indiana State University, Indiana, Recorder

Robert Daiber, Triad High, Illinois

Jeffrey Grimmer, Mankato East High, Minnesota

Norman Hackerman, The Robert A. Welch Foundation, Texas

Michael Jensen, Paonia High, Colorado

Michael Mino, The Gilbert School, Connecticut

Scott Warner, Lawrenceburg High, Indiana

George Willcox, Virginia Department of Education, Virginia

Representatives

Carl Hall, National Academy of Engineering, Washington, D.C.

Flint Wild, National Aeronautics and Space Administration, Washington, D.C.

Advisory Group

Rodger Bybee, Executive Director, Center for Science, Mathematics, and Engineering Education, National Research Council, Washington, D.C.

Daniel Goroff, Mathematics Department, Harvard University, Massachusetts

Thomas Hughes, Jr., Director of Development, Foundation for Technology Education, Virginia

George Nelson, Director, Project 2061, American Association for the Advancement of Science, Washington, D.C.

James Rutherford, Education Advisor, Project 2061, American Association for the Advancement of Science, Washington, D.C.

Linda Rosen, Former Executive Director, National Council of Teachers of Mathematics, Virginia

Kendall Starkweather, Executive Director, International Technology Education Association, Virginia

Gerald Wheeler, Executive Director, National Science Teachers Association, Virginia

William Wulf, President, National Academy of Engineering, Washington, D.C.

National Research Council's Standards Review Committee

William A. Wulf, National Academy of Engineering, Washington, D.C., Chair

Karin Borgh, BioPharmaceutical Technology Center Institute, Madison, Wisconsin

Rodger Bybee, Biological Sciences Curriculum Study (BSCS), Colorado Springs

Elsa Garmire, Dartmouth College, Hanover, New Hampshire

James Rutherford, American Association for the Advancement of Science, Washington, D.C.

National Academy of Engineering Focus Group Review

Alice Agogino, Professor, University of California

George Bugliarello, Chancellor, Polytechnic University, New York

Samuel Florman, Chairman, Kreisler Borg Florman Construction Company, New York

Elsa Garmire, Professor, Dartmouth College, New Hampshire

Carl Hall, Engineer, Engineering Information Services, Virginia

John Truxal, Professor, State University of New York at Stony Brook

National Academy of Engineering Special Review Committee

Richard Alkire, University of Illinois at Urbana-Champaign

Frank Aplan, Pennsylvania State University

Shu Chien, University of California, San Diego

George Fox, Grow Tunneling Corporation, New York

William Friend, Bechtel Group, Inc., Washington, D.C.

Elmer Gaden, Jr., University of Virginia

Donald Johnson, Grain Processing Corporation, Iowa

Dean Kamen, DEKA Research and Development Corporation, New Hampshire

David Kingery, University of Arizona, Arizona

John Lee, Texas A&M University, Texas

Margaret LeMone, National Center for Atmospheric Research, Colorado

Matthys Levy, Weidlinger Associates, New York

Henry Paynter, Massachusetts Institute of Technology, Massachusetts

Greg Pearson, National Academy of Engineering, Washington, D.C.

Jerome Schultz, University of Pittsburgh, Pennsylvania

Hardy Trolander, The Yellow Springs Instrument Company, Inc., Ohio

National Research Council's Technical Review Panelists

Dennis Cheek, Rhode Island Department of Education, Rhode Island

Rodney Custer, DTE, Illinois State University, Illinois

Denny Davis, Washington State University, Washington

Marie Hoepfl, Appalachian State University, North Carolina

Larry Leifer, Stanford University, California

Peggy Lemone, University Corporation for Atmospheric Research, Colorado

Thomas Liao, SUNY at Stony Brook, New York

Franzie Loepp, DTE, Illinois State University, Illinois

A. Frank Mayadas, Alfred P. Sloan Foundation, New York, New York

Bruce Montgomery, MIT/PSFC, Cambridge, Massachusetts

John Ritz, Old Dominion University, Norfolk, Virginia

Mark Sanders, Virginia Tech, Blacksburg, Virginia

Fredrick Stein, CSMATE, Fort Collins, Colorado

George Toye, wiTHinc Inc. and Stanford University Learning Center, California

Scott Warner, Lawrenceburg High School, Indiana

Kenneth Welty, University of Wisconsin-Stout, Wisconsin

Jane Wheeler, Monte Vista Elementary School, Rohnert Park, California

The National Commission for Technology Education (Phase I, 1994-96)

G. Eugene Martin, School of Applied Arts and Technology, Southwest Texas State University, Texas, Chairperson

J. Myron Atkin, Stanford University, California

E. Allen Bame, Virginia Tech, Virginia

M. James Bensen, DTE, Bemidji State University, Minnesota

Gene R. Carter, Association for Supervision and Curriculum Development, Washington, D.C.

Robert A. Daiber, Triad High School, Illinois

James E. Davis, Ohio University, Ohio

Paul W. Devore, DTE, PWD Associates, West Virginia

Ismael Diaz, Fordham University, New York

William E. Dugger, Jr., DTE, Technology for All Americans Project, Project Director, Virginia

Frank L. Huband, The American Society for Engineering Education, Washington, D.C.

Thomas A. Hughes, Jr., Foundation for Technology Education, Virginia

Patricia A. Hutchinson, Trenton State College, New Jersey

Thomas T. Liao, Sate University of New York at Stony Brook, New York

Franzie L. Loepp, Center for Mathematics, Science, and Technology, Illinois

Elizabeth D. Phillips, Michigan Sate University, Michigan

Charles A. Pinder, Northern Kentucky University, Kentucky

William S. Pretzer, Henry Ford Museum and Greenfield Village, Michigan

John M. Ritz, DTE, Old Dominion University, Virginia

Richard E. Satchwell, Technology for All Americans Project, Virginia

Kendall N. Starkweather, DTE, International Technology Education Association, Virginia

Charles E. Vela, MITRE Corporation, Virginia

Walter B. Waetjen, Cleveland State University, Ohio

John G. Wirt, Columbia University, New York

Michael D. Wright, University of Missouri-Columbia, Missouri

Field Review Sites

Agawam Junior High School, Massachusetts—John Burns, James Graveline, and David Littlewood

Agawam Middle School, Massachusetts—Maynard Baker

Belleview High School, Florida—Dale Toney

Bloomfield Hills Middle School, Michigan—Al Binkholz

Burris Laboratory School, Indiana—James Kirkwood, DTE

*Cass Elementary School, Michigan—James Lauer

Caverna Junior/Senior High, Kentucky—Dennis Bledsoe

Century High School, Oregon—Chip Miller

Columbia City Elementary School, Florida—Cheryl Cox

Cumberland Valley High, Pennsylvania—Robert Rudolph

Cutler Ridge Middle School, Florida—Jeff Meide

Damascus High School, Maryland—Robert Eilers, Robert Turnbull, and George Thomas

Dan River High School, Virginia—Robert Huffman

Dublin Scioto High School, Ohio—Kevin Burns

Forest Hills High School, Pennsylvania—Terry Crissey

Fruita Monument High School, Colorado—Ed Reed

G. Ray Bodley High School, New York—Tom Frawley

Gilbert School, Connecticut—Mellissa Ann Morrow

Gladstone High School, Oregon—Roy DeRousie

Greenfield-Central High School, Indiana—Gary Wynn, DTE

*Grant Wood Elementary, Iowa—Sandra Ann Lawrence

Harrisonville Middle and High, Missouri—David Vignery

*Hellgate Elementary School, Montana—Bruce Whitehead

Hermosa Valley School, California—Teri Tsosie

Hershey Middle School, Pennsylvania—Kevin Stover

Hobart Middle School, Indiana—Bob Galliher

Hoffman T.E.C.H. Center, Illinois—Cecil Miller

Homewood High School, Alabama—Leah Griffies

John T. Baker Middle School, Maryland—Brian Niekamp

Kalida High School, Ohio—Dale Liebrecht

LakeVille High School, Michigan—Dennis Harrand

*Lange Middle School, Missouri—Carole Kennedy

Lehman Middle School, Ohio—John Emmons

Louis M. Klein Middle School, New York—Henry Strada

Marlboro Middle School, New Jersey—Alan Lang

Monte Vista Elementary School, California—Jane Wheeler, Carey Dahlstrom, Diana Klein, Betsy Smith, and Carol Licht

Normal Community West High, Illinois—Jim Boswell

Pine River Area, Michigan—Carla Chaponis, Mike Maskill, and Carol Posey

Pleasant Hill Elementary School, Texas—Vanessa Jones

Rancho Cotate High School, California—Adam Littlefield

Rochelle Township High School, Illinois—Rick Bunton

Rocky Hill Middle School, Maryland—Michael Callaway

Suncoast Elementary School, Florida—Jo Ann Hartge

Syosset High School, New York—Barry Borakove

Thomas Dale High School, Virginia—Christopher Kelly

Twelve Corners Middle School, New York—Joseph Priola

Venice High School, Florida—Arnall Cox

Vernon Township Public Schools, New Jersey—Mark Wallace, Gaylon Powell

Westlake High School, Ohio—Scott Kutz

*Representatives of National Association of Elementary School Principals (NAESP)

ITEA and TfAAP wish to thank the National Science Foundation (NSF) and the National Aeronautics and Space Administration (NASA) for their funding of Phase II of the project. Special appreciation is given to Gerhard Salinger of NSF and Frank Owens of NASA for their advice and input.

We would like to thank William Wulf, President of the National Academy of Engineering, who chaired the National Research Council Standards Review Committee and the National Academy of Engineering Focus Group which reviewed the final drafts of *Technology Content Standards*. Also, we would like to thank Greg Pearson, who coordinated the National Academy of Engineering Focus Group Review and Special Review Committee. We appreciate the advice and council from Pamela Mountjoy and Flint Wild of NASA. A special thanks to Fred Brown and Gail Connelly Gross of the National Association of Elementary School Principals for their involvement in the field review of the document.

ITEA and TfAAP also would like to express our appreciation to Jack Frymier and Jill Russell, representatives of Phi Delta Kappa, for serving as our evaluators throughout Phase II.

The project staff would like to thank Robert Pool for his expertise and creativity in writing and editing sections of the book. We are also appreciative of his willingness to attend numerous development meetings in order to gain a greater understanding of the subject matter and of the project's vision.

Regarding Draft 5, special thanks is extended to John Wells for expertise and suggestions on incorporating elements of his Technology Education Biotechnology Curriculum (TEBC) taxonomy. Also, we thank selected members of the NRC Technical Review Committee for their editorial and content input.

Thanks also to John McCormick for taking photographs, Ed Scott and the staff at Harlow Typography for designing the book, RR Donnelley & Sons Company for printing the standards. Also we thank Rhonda Simmerman, Stephanie Overton, and Brigitte Valesey for their work in editing the standards, and Jeff Swab for help in designing our homepage. Special appreciation is given to Jack Hehn, manager, Education Division, American Institute of Physics for his review and edit of Standard 16, Energy and Power Technologies. Also we would like to thank Bob Englander of Englander Indexing Services for creating the index. Thanks to Deanna Colaianne for help in the final editing of the document.

Additionally, we wish to acknowledge the TfAAP staff, who generously supported each step of the development process. Also, thanks to Jodie Altice, Diane Kitts, Amy Mearkle, and Taryn Sims, former staff members, for their enthusiasm and devotion to the project.

Reviewers

We would like to express appreciation to all of the reviewers of *Technology Content Standards*. This list includes people from around the world who provided valuable input into strengthening the document. There are many different backgrounds represented in this group of reviewers who came from the ranks of elementary and secondary teachers; supervisors at the local and state level; teacher educators; administrators including elementary and secondary school principals; school system superintendents; engineers; and scientists; parents; school board members; international leaders in technology education; and others. *Technology Content Standards* has been improved in its various iterations as a result of the comments and recommendations of all reviewers.

Gary Aardappel
Patrick Abair
George Abel
Paul Adam
Michael Adams
Stephanie Adams
Alice Agogino
Paul Agosta
Shaul Aharoni
John Aiken
Tom Akins
Ari Vilppu Alamaki
Bryan Albrecht
Nelle Alexander
Donald Allaman
Cynthia Allen
Jerry Allen
Paul Allen
Jay Alsdorf
Richard Ambacher
Alex Amoruso
Kenneth Amos
Darrell Andelin
Byron Anderson
Debbie Anderson
Sheila Anderson
Ernie Ando
Ronald Ankeny
Piet Ankiewicz
Frank Aplan
Mike Appel
Richard Archibald
Spence Armstrong
Wayne Arndt
Steven Ashby
Pamela Askeland
Denise Atkinson-Shorey
Robert Austin
Dwight Back
Bruce Baker
Richard Baker
Maynard Baker
Jerry Balistreri, DTE
William Ball, DTE
Allen Bame
Mohamed Bandok
Victoria Bannan
Gary Bannister
Moshe Barak
Steven Barbato
Ronald Barker
Colleen Barnes
Michael Barnes
Wiley Barnes
Bill Barowy
Michael Barry
Tom Barve
Lynn Basham
Susan Bastion
Michael Bastoni
Debra Baxter
Steven Baylor
Jon Bean
Michael Beck
John Beggs
Mark Beise
Thomas Bell
Gary Bender
Christine Bengston
Barbara Bennett
Russell Bennett
James Bensen
Clare Benson
David Beranek
Michael Beranek
Richard Bergacs
William Berggren
Gordon Bernard
Kenneth Bertrand
Julia Best
Carl Betcher
Bill Bewley
William Bien
David Biggs
Keith Bigsby
David Billington
Ken Bingman
Al Binkholz
Burton Bjorn
Gerry Blackburn
Crystal Blackman
Dennis Bledsoe
Cynthia Blodgett-McDeavitt
Roy Blom
Carla Boeckman
Dennis Bohmont
Jill Bohn
Kelly Bolender
Janet Boltjes
Dominick Bonanno
Kathleen Bond
Paul Bond
Ernest Bonner
Wayne Bonsell
Lewis Boone
R. J. Booth
Barry Borakove
Karin Borgh
Jim Boswell
John Brandt Botti
Charles Boucher
Roy Boudreau
Paul Bouffard
David Bouvier
Ramona Bowers
Sue Boyer
Dianne Braden
Morgan Branch
Chad Brecke
Andy Breckon
Damuan Breeze
Norman Brehm
Bruce Breilein
Kenneth Bremer
Gail Breslauer
Carole Briggs
Gregory Briggs
Andrew Britten
Edward Britton
Katherine Brophy
John Brown, DTE
Paul Brown
Ron Brown
Timothy Brown
Tyson Brown
Alan Brumbaugh
Silas Bruner
Sharon Brusic
George Bugliarello
Margery Brutscher-Collins
Leah Bug-Townsend
Michael Bunner
Rick Bunton
Walter Burgin
Barry Burke, DTE
Jerry Burmeister
John Burns
Edward Burton
Joe Busby
Jeffrey Bush
Edward Butler
James Butler
Peter Butler
Keith Butterfield
Rodger Bybee
Fernando Cajas
Anthony Calabrese
Richard Call
Kristin Callender
Michael Callaway
Allan Cameron
Anne Campbell
Brian Canavan
Roger Cantor
Hershey Card
Vaughn Cardashian
Phillip Cardon
David Carey
Mary Agnes Carling
Patrick Carlson
Christopher Carroll
Del Carson
Frank Casey
James Cedel
Alverna Champion
Susan Chandler
Ken Chapman
David Chatland
Victor Chavez
John Chen
Shu Chien
Vincent Childress
Dean Christensen
Jim Christensen
Gary Christopher
Karen Christopherson
Thomas Claassen
Craig Clark, DTE
David Clarke
Barbara Clements
Toss Cline
Beatrice Clink
Margaret Clinton
Sam Cobbins, DTE
Kenneth Cody
Alden Colby
Mark Coleman
Kathryn Collins
Kenneth Collins
Sharon Collins
Lizabeth Comer
Douglas Cone
Paul Contreras
Bruce Coon
Judith Cope
Charles Corley, DTE
William Cosenza, Jr.
Joel Cotton
Sam Cotton
Ed Coughlin
Bryce Coulter
Peggy Cowan
Arnall Cox
Barry Cox
Cheryl Cox
Alan Cram
Caren Cranston
Terry Crissey
Pamela Croft
Lois Ann Cromwell
Terry Cross
Larry Crowder
Anita Cruikshank
David Culver
Roy Culver
Barbara Cummins
Glenn Current
Rodney Custer, DTE
Jennifer Cutler
F. J. Cutting
Christopher Cytera
Osnat Dagan
Richard Dahl
John Dahlgren
Robert Daiber
Richard Daignault
Wayne Dallas
Thomas D'Apolito, DTE
Donald Darrow
Michael Daugherty
Jack Davidson
Kevin Davis
Richard Davis
Phillip Dean
Robert Deans
Brad Dearing
Beverly DeGraw
Greg Dehli-Young
Michael De Miranda
Carol Denicole
Clair Denlinger
Denise Denton

Ed Denton
Howard Denton
Peter Denzin
Roy DeRousie
David Devier, DTE
Thomas Devlin
Paul DeVore
Marc de Vries
Yasin Dhulkifl
Ismael Diaz
Dan Dick
Eric Dickie
James Dieringer
Doug Dillion
Randy Dipner
Anthony Docal
James Dolan
Spencer Dolloff
Tim Dolphay
Jon d'Ombrain
Lori Donnelson
Robert Dorn
Michael Doyle
David Drexter
Mike DuBois
Kenneth Dues
Frank Duggan
Stephen Duncan
Larry Dunekack
Robert Dunkle
Dorothy Dunn
Robert Dunn
Ron Dunn
Phyllis Dunsay
Phyllis Durden
Bill Dutton
Nancy Dyck
Michael Dyrenfurth
Vanik Eaddy
Naomi Edelson
Glenn Edmison, DTE
Robert Eilers
Mauka Elem
Jeffrey Elkner
Deborah Elliott
Audie Ellis
Nathaniel Ellis
Nancy Elnor
Leo Elshof
Mark Emery
John Emmons
Sherman England
Dan Engstrom
Ronald Engstrom
Julie Enstrom
Thomas Erekson
David Erlandson
Don Eshelby
Neil Eshelman
Rosemary Eskridge
Bruce Evans
Mark Evans
Marsh Faber
Leonard Fallscheer
Eloise Farmer
Geraldine Farmer
John Fecik
Paul Fecke
Sheryl Feinstein
Esperznaz Fernandez
John Fialko
William Finer
Craig Firmender
Ronald Fisher
John Fitzgibbon
James Fitzpatrick
Stephen Florence
Samuel Florman
Judy Forman
Richard Foster
Tad Foster
Gary Foveaux
David Fowler
Dave Frankmore
David Fraser, DTE
Tom Frawley
Boe Fred
Kathy Fredrick
Richard Freeburg
Brad Freeman
William Friend
David Fromal
Bruce Fuchs
Sherry Fuller
Elmer Gaden, Jr.
Michael Gadler
Thomas Gaffney
Sue Galayda
Robert Galliher
Carl Gamba
Joyce Gardner
Mike Gargiulo
Elsa Garmire
Marlane Garner
Jennifer Gasser
Cyndi Gaudet
Brett Gehrke
Peter Genereaux
Bradford George
James Gertz
Elissa Gerzog
Don Getzug
James Gianoli
Anthony Gilberti
Doran Gillie
Dorothy Gleason
Richard Glueck
Darlene Godfrey
Adele Gomez
Tom Good
Linda Goodwin
Morris Gordon
Dave Gorham
Kenneth Gornto
Daniel Goroff
Kevin Gotts
Manny Grace
Gary Graff
Sandi Graff
Kathrine Graham
Thomas Gramza
Laury Grant
Wayne Grant
James Graveline
Donald Gray
Joy Gray
John Gray, DTE
Clark Greene
David Greer, DTE
Laura Grier
Leah Griffies
Jeff Greuel
Ann Grimm
Jeffrey Grimmer
Richard Grimsley
June Grivetti
Eric Gromley
Marvin Grossman
Peter Grover
Mae Groves
Darren Grumbine
Karen Guillet
Genesta Guirty
Jack Gundolfi
Jean Guzek
Michael Hacker
Norman Hackerman
Carl Hader
Jaonne Hagedorn
Cindy Hager
Carl Hall
Jack Hall
John Hall
Linda Hallenbeck
Christopher Halloran
Dale Hanson
Robert Hanson
Kevin Hardy
Lawrence Hardy
Paul Harley
Henry Harms
Andrea Harpine
Carroll Harr
Dennis Harrand
Elden Harris
Myril Harrison
Margaret Harsch
Jo Ann Hartge
Helen Hasegawa
Dean Hauenstein
James Havelka
Brad Hawk
Shirley Hawk
Lynn Hawkins
Simon Hawkins
Tina Hayden
Daryl Hayes
Douglas Hayhoe
William Haynie, III
Irene Hays
Carl Healer
Don Healer
Roland Hebert
Chad Heidorn
Neil Heimburge
Jan Heinrich
Hal Helsley
Merritt Hemenway
Joyce Henstrand
Tom Hession
Clynell Hibbs
Garth Hill
Gerald Hill
Jane Hill
Mike Hill
Robert Hinderer
Lee Hipkiss
Linda Hodge
Marie Hoepfl
Lorna Hofer
Donald Hoff
William Hoffman
Anne Holbrook
Jim Holderfield
Harold Holley
Sidney Holodnick
Elroy Holsopple
Michael Hoots
Gerd Hopken
Alan Horowitz
Daniel Householder, DTE
Fred How
Mark Howard
John Howarth
Tim Howes
James Howlett
Michael Hoye
Hai Hu
Robert Huffman
Thomas Hughes, Jr.
Van Hughes
Dale Hummel
Damon Hummel
Susan Huntleigh-Smith
William Husby
Patricia Hutchinson
Joseph Huttlin
Steve Ickes
Nicholas Iliadis
Anthony Infranco
Everett Israel, DTE
Michael Ive
Michael Izenson
David Izzo
Ron Jacobitz
Joe Jakopic
Lee James
David Jarzabek
Judy Jeffrey
Thomas Jeffrey
Susan Jeffries
Gerald Jennings
Mike Jensen
Lars Jenssen
Jeffrey Jobst
Hardy John
Albert Johnson
Arden Johnson
Arthur Johnson
Bud Johnson
Donald Johnson
Glenn Johnson
Jimmie Johnson
Ken Johnson
Maren Johnson
Richard Johnson
Todd Johnson
Alister Jones
Docia Jones
John Jones
Mark Jones
Patricia Jones
Vanessa Jones
Kathleen Jordan
Joe Josephs
Richard Junkins
James Justice
Roberta Kaar
Ronald Kahn
Stephen Kahn
Marie Kaigler
Lee Kallstrom
Dean Kamen
Tapani Kananoja
Gregory Kane
James Kane
Anne Kanies
Rolland Karlin
Robert Karolyi
John Karsnitz
Wayne Kazmierczak
Philip Keefer
Rob Keeney
Pat Keig

Ed Keller
Margaret Keller
Christopher Kelly
Colleen Keltos
Michale Kemp
James Kendrick
Carole Kennedy
James Kennedy
Kenneth Kern
Wanjala Kerre
Scott Kessler
Scott Kiesel
Richard Kimbell
Rene Kimura
Charlene Kincaid
Brent Kindred
Cyril King
David King
David Kingery
Mary Kinnick
James Kirkwood, DTE
Kurt Klefisch
Glenn Klutz
James Knapp
Zondra Knapp
Stephan Knobloch
Steve Knox
Judith Koenig
Matthew Koliba
Stan Komacek
Kevin Konkel
Holly Koon
Stephen Koontz
Curt Kornhaus
Susan Kostuch
Denny Kozita
John Kraljic
Phillip Krueger
Gerald Kuhn
M. Kunesh
George Kunkle
Scott Kutz
Paul Kynerd
Judy Lachvayder
Henry Lacy
John Laffey
Teri Lampkins
Wayne Lancaster
Patricia Lancos
Tabitha Landis
C. Landry
Alan Lang
Wayne Lang
James LaPorte
Ted Larsen
Connie Larson
Gini Larson
Victor Larson
Stanley Lathrop
James Lauer
Keith Lavin
Christopher Lawn
Sandra Ann Lawrence
M. Lazaraton
Scott LeCrone
John Ledgerwood
Brett Lee
Evelyn Lee
Hillary Lee
John Lee
Lung-Sheng Lee
Mark Leeper
Amanda Leinhos
Margaret LeMone
Scott Lenz
Victor Leonov
Bob Lesch
James Levande
Jane Leven Cole
Walter Lewandowski
Fred Lewis
Jerry Lewis
Peter Lewis
Theodore Lewis
Thomas Liao
Barbara Libby
Dale Liebrecht
Bill Lind
Leah Lindblom
Robert Lindemann
Sue Linder
William Linder-Scholer
Theodore Lindquist
Kerry Lintner
Len Litowitz, DTE
David Littlewood
Chin-Tang Liu
Brent Lockliear
Franzie Loepp, DTE
Kevin Logan
Deborah Longeddy
Roger Lord
Susan Loucks-Horsley
Gerald Lovedahl, DTE
Diane Lovins
Peter Lowe
Phillip Loyd
Steven Lubar
Judith Luce
Laura Lull
Charles Lupinek
Pierre Lussier
Robert Lyle
James Lynch
Kathleen MacNaughton
Lee Madrid
Tara Magi
G. S. Mailhot, Sr.
T. Majors
Norb Malik
Chiquita Marbury
Roger Marchand
Linda Markert
Suzanne Marks
Lynn Marra
Benjamin Martin
Corrie Martin
Gary Martin
Gene Martin
John Martin
Pete Martin
Mike Maskill
Katharine Mason
Joanne Masone
Janne Mathes
Jill Mathes Prokop
Nancy Mathras
Edward Mattner
Kathleen Mau
Mike Maxwell
Brian McAlister
Joseph McCade
Tim McCarty
Gerald McConaghy
Robert McCormick
Ann McCoy
David McCready
David McCrory
Douglas McCue
Linda McElvenny
Phillip McEndree
Dennis McGowan
Kelli McGregor
Randy McGriff
Robert McGruder
Robert McIntosh
Heather McKenna
Stephen McKenzie
Charles McLaughlin
Maggie McLean
Tim McNamara
Sue McPherson
Mark McVicker
Steve Megna
Jeff Meide
Martin Meier
Sherry Meier
Kathleen Melander
Darcy Mellinger
Frank Meoli
Joseph Merenda, Jr.
LaVon Merkel
Jody Messinger
Howard Middleton
Dan Migliorini
John Mihaloew
Al Miller
Chip Miller
Dyann Miller
Kenneth Miller
Kevin Miller
Michael Miller
Russell Miller
Simon Miller
Stanley Miller
Kevin Milner
Charles Minear
Michael Mino
Andrew Mitchell
Sharon Miya
Hidetoshi Miyakawa
J. P. Mobley
Annette Moders
Torin Monahan
Robert Montesano
Perry Montoya
Kevin Moorhead
Bob Morris
Linda Morris
Laura Morrison
Mellissa Ann Morrow
Tim Motherhead
Denis Mudderman
Sharon Muenchow
Jamie Mulligan
Clint Mullins
Shoji Murata
Gregory Murphy
Harrison Murphy
Janet Murphy
Don Musick
Veny Musumecci
Donald Mydzian
James Myers
Ilia Natali
Terry Neddenriep
Michael Neden
Diane Neicheril
Mark Nellkam
George Nelson
Mary Nemesh
Albert Newberry
Peter Newell
Gerri Newnum
Richard Nicholson
Brian Niekamp
Chris Nielsen
Ken Nimchuk
Ben Ninoke
Scott Noles
Lee Noonan
Gerald Nordstrom
Hana Novakova
James Novotny
Brian Nulsan
J. Nuzzo
Kathaleen O'Bosky
Karen O'Donovan
Steven O'Green
Tom Oahs
Lynn Ochs
Tom Ochs
Virginia Okamoto
Nancy Oppenlander
Susan Oppliger
William Oppliger
Paul Oravetz
Richard Ortega
Donna Ostrowski-Cooley
Deborah Owens
Clarence Owens, Sr.
Ned Owings
Brad Paddock
Beverly Paeth
Brenda Page
Cathy Bradley Page
William Paige, DTE
Russell Palumbo
Ellen Pantazis
Scott Papenfus
Berlin Parker
Jill Parker
Teresa Parks
James Partridge
Paul Parzych
Gale Passenier
Lin Patty
Alan Paul
Alona Paydon
Judy Payne
Henry Paynter
Mary Pearce
Diana Pearson
Greg Pearson
Peter Pedersen
Barbara Pellegrini
Michael Pennick
William Pennington
Marjorie Pentoney
Don Perkins
Todd Perkins
Charles Perrin
James Perryman
Wesley Perusek
Frank Pesce
Donald Peterson
Michael Peterson
Bruce Peto
Stephen Petrina
Edward Pfeifer
Betty Phillips
Dennis Phillips
Kenneth Phillips
David Pickhardt
Thomas Pieratt

Alan Pierce
Randal Pierce
Steven Pille
David Pilo
Charles Pinder
Liliane Pintar
Theodore Piwowar
Kelly Podzimek
Beth Politz
Gary Porter
Carol Posey
Paul Post
Andy Potter
Richard Potts
Gaylon Powell
J. Powell
Bill Powers
Eldon Prawl
William Pretzer
Hugh Price
Joseph Priola
Kurt John Proctor
Dennis Ptak
Annie Purtill
Katrina Pyatt
David Quinn
Anna Quinzio-Zafran
Sidney Rader
Daniel Raether
William Ragiel
Senta Raizen
Richard Ralstin
Gene Ranger
Gregory Ratliff
Robert Raudebaugh
Martin Reardon
David Redding
Roderick Reece
Ed Reed
Julene Reed
Philip Reed
Edward Reeve
David Reeves
James Regilski
JoAnne Reid
Rondel Reynolds
Jeff Rhodus
Michael Ribelin
Mark Richardson
Barbara Riester
Diana Rigden
Marian Rippy
Patricia Ritchey
John Ritz
David Roae
Donna Roberson
Jessie Roberts
Roger Robidoux
Marsha Robison
George Rockhold
Rose Rodd
Mark Rodriguez
Nelson Rodriquez
John Roeder
Steve Rogers
Harry Roman
Ruben Romero
Douglas Root
Mary Rose
Linda Rosen
Becky Ross
Raymond Ross
Jerry Roth
Zipora Roth
Christopher Rowe
Rob Roy
Ruth Rozen
Robert Rudolph
James Rutherford
Ernest Ruiz
John Ryden
Betsy Rymes
Matt Sabin
Marsha Sagmoe
Nicasio Salerno
Esteban Salinas
Gerhard Salinger
Frank San Felice
Vern Sandberg
Mark Sanders
Karen Sanko
Richard Sanko
Gary Santin
Scott Sassaman
Richard Satchwell
B. D. Satterthwaite
Ernest Savage,
DTE
Randy Schaeffer
Barry Schartz
Arthur Schattle
Glenn Schenenga
Jo Schiffbauer
K. Schipper
Brian Schmidt
Rick Schmidt
Diane Schmidtke
Laurie Schmitt
Jane Schmottlach
Tonia Schofield
Frederick Schroedl
Jerome Schultz
Anthony Schwaller,
DTE
Robert Scidmore
Mark Sebek
Debra Seeley
David Seidel
Marvin Selnes
Bolivar Senior
Marty Sexton
Robert Sexton
Ray Shackelford
Rich Shadrin
Richard Shadrin
Don Shalvey
Theodore Shatagin
Kathy Sheehan
Jeenson Sheen
John Sheley
Leonard Shepherd
Rick Shepker
Mark Sherman
George Shield
Mary Shield
Geoffrey Shilleto
Scott Shook
Thomas Shown
Deborah Shumate-Wesbrook
Matt Sinclair
Ron Skalsky
Dennis Skurulsky
Jeffrey Smith
Robert Smith
Rus Smith
Terrel Smith
Theresa Smith
Thomas Smith
Woodrow Smith
Bart Smoot
William Snelson
Kyo Song
Valerie Sorensen
Lawrence Soscia
Linda Southworth
Joseph Spadavecchia
Loretta Speed
Ken Spellman
John Spencer
Mark Spoerk
Kevin Squires
Kay Stables
Cathy Stacy
Richard Stacy
Craig Stahly
Jan Stark
Ken Starkman
Kendall Starkweather
Gregg Steele, DTE
Nicholas Steill
Sam Steindel
Paul Stengel
Katherine Stephens
Leonard Sterry
Mike Stevens
Gary Stewardson
Lori Stewart
Howard Stob
Ken Stovall
Kevin Stover
Scott Stowell
Henry Strada
Nancy Strada
Jason Strate
Beth Stroh
Eric Suhr
Gregory Sullivan
Kazuhiro Sumi
Gary Surratt
Kris Swanson
Darlina Swartz
Rick Swartz
Andrew Sweeney
Neal Swernofsky
John Sylvester
Mary Szoka
Judy Tamfu
Ed Taylor
Rick Taylor
Terrance Taylor
Becki Teague
Jim Teicher
Jeffrey Testa
Paul Thallner
Charles Theis
George Thomas
John Thomas
Daniel Thompson
Duren Thompson
Eric Thompson
Kristy Thompson
Steven Thompson
Carole Thomson
Charles Thorneycroft
Kathy Thornton
Kathy Tibone
Wade Tischner
Barbara Todd
Donald Todd
Ronald Todd
Norman Tomazic
Dale Toney
Steven Topp
Sherri Torkelson
Dorothy Towler
Alicia Townsend
Bernie Trilling
Leon Trilling
Hardy Trolander
Dan Troshynski
Dan Troxel
Philip Trudeau
John Truxal
Teri Tsosie
Dave Tundo
Robert Turnbull
Bill Turner
Hank Turner
William Turner
Matthew Uibel
Scott Underwood
Brian Uslan
Charles Utz
John Vaglia
Richard Valencia
Edward Valentukonis
Joyce Valenza
Linda Valenzuela
Brigitte Valesey, DTE
Dean Vanderbyl
Eric Van Duzer
Arvid Van Dyke
Paul Van Hulle
Carol VanSpybrook
Marguerite Vavalla
Steve Vergara
Margaret Vescio
Paul Vetrano
David Vignery
Chris Vodopich
Daniel Vrudny
Scott Vugteveen
Walter Waetjen
Doug Wagner
Vincent Walencik
Carmen Walker
Terry Walker
Mark Wallace
Gregg Walls
Kathleen Walters
Cathy Walton
Kin Kwok Wan
Changhua Wang
Daniel Wang
Ding Ming Wang
Michael Warner
Scott Warner
Shelly Wasson-McRel
Peggy Watson
Gina Webber
Brian Webberley
Robert Weber
Jerry Weddle
Paul Weir
Barry Weisberg
Cheryl Welch
Malcolm Welch
Sandra Wellens
Chris Weller
Jack Wellman
John Wells
Ken Welty
Rob Weneck
Daniel Weselak
Geoffrey Westervelt

Gerald Wheeler
Jane Wheeler
Mary White
Bill Whited
Bruce Whitehead
Petrina Whiteside
Jack Whiting
Lorraine Whitman
Mike Whittaker
Skip Wiarda
Robert Wicklein, DTE
Chris Widmer
Sandy Wiegers
Emerson Wiens
Flint Wild
Jim Wiljanen
George Willcox
Darrell Williams
Deborah Williams
John Williams
Marcia Williams
David Wilson
Mark Wilson
Melissa Wing-Ronca
George Winner
John Wirt
Jane Wisniewski
Kendra Wofford
Leonard Wolf
Chris Wonderly
Bobby Woods
Barbara Worden
Wes Worley
Charles Wright
John Wright
Michael Wright, DTE
Thomas Wright, DTE
William Wulf
Connie Wyant
Mary Wyatt
Gary Wynn, DTE
Rusty Wynn
John Wyrick
Dorothy Yager
Joseph Yarbrough
Connie Yeatts
Wallace Yoho, DTE
Deanna Young
Posey Young
Robert Young
LaVerne Young-Hawkins
Chien Yu
Ronald Yuill, DTE
John Zahn
Edward Zak
Ron Zanini
Kimberly Zeidler
Eric Zelanko
Pamela Zelaya
Fredele Zouzounis
Jim Zucchetti
Karen Zuga

Vignette Credits

page 29. *Creating a safer working environment.* Adapted from a vignette written by James Graveline, Agawam Junior High School.

page 37. *The bicycle as a vehicle for learning.* Written by Sharon Brusic, Virginia Polytechnic Institute and State University.

page 47. *Helping out Stuart Little.* Adapted from a Project UPDATE vignette.

page 53. *A hands-on experience.* Adapted from a vignette written by Sharon Brusic, Virginia Polytechnic Institute and State University.

page 64. *Students plan new airport site.* Written by Sharon Brusic, Virginia Polytechnic Institute and State University.

page 70. *The best bag in Agawam.* Adapted from a vignette written by John Burns, Agawam Junior High School.

page 75. *The development of the button.* Adapted from a vignette written by Sharon Brusic, Virginia Polytechnic Institute and State University.

page 82. *A time line comparison of communicating a message.* Written by Catherine Ney, Christiansburg Middle School.

page 96. *Designing a gift of appreciation.* Adapted from a vignette written by Sharon Brusic, Virginia Polytechnic Institute and State University.

page 101. *Can you help Mike Mulligan?* Adapted from a vignette written by Marie Kellam-Cook and Amy Hackett, students of James Kirkwood, Ball State University.

page 109. *Navigational technology.* Adapted from a vignette written by David Bixby and Bruce Whitehead, Hellgate Elementary School.

page 117. *Building something to float.* Adapted from a vignette written by Carole Thomson, Northern College, Scotland.

page 122. *The great paper car race.* Adapted from a vignette written by Bob Galliher, Hobart Middle School.

page 125. *The America's cup challenge.* Written by Rob Cronk, reprinted with permission from International Technology Education Association. (1995). Technology Learning Activities II. Reston, VA: International Technology Education Association, 22.

page 129. *Take it apart.* Reprinted with permission from Wells, J. G. and S. A. Brusic. (1993). Mission 21: Launching technology across the curriculum. New York: Delmar Publishers, Inc., 128-131.

page 136. *Clean up an oil spill.* Adapted from a vignette written by Tiffany Cupp, Cassie Fugiett, Sarah Meyers, and Jenny Shumowsky students of James Kirkwood, Ball State University.

page 144. *A Pharmacy connection.* Suggested by J.G. Wells, West Virginia University.

page 157. *Hydroponics system.* Written by Franzie Loepp, DTE, Illinois State University.

page 161. *Developing and producing a product or system collaboratively.* Adapted from the work of Linda Gostomski and Pat Ward-Mytinger, Roosevelt Elementary School.

page 172. *Communication through a home page on the web.* Adapted from a vignette written by Scott Kutz, Westlake High School.

page 188. *A team approach to plastics.* Adapted from a vignette written by Sharon Brusic, Virginia Polytechnic Institute and State University.

page 197. *A look at energy efficient homes.* Adapted from a vignette written by Scott Kutz, Westlake High School.

page 215-219. *Articulated curriculum example from Grades K-12.* Written by Sharon Brusic and Mark Sanders, Virginia Polytechnic Institute and State University.

Glossary

The terms defined and described in this glossary apply specifically to the *Technology Content Standards.* These terms may have broader meanings in different contexts.

Agriculture — The raising of crops and animals for food, feed, fiber, fuel, or other useful products.

Agroforestry — Land management for the simultaneous production of food, crops, and trees or the intentional designing of land through a system of planting trees, shrubs, crops, or forage in order to improve habitat values, access by humans and wildlife, and woody plant products.

Alternative energy source — Any sources or resources of energy that are renewable through natural processes, can be renewed artificially, or that are regarded as practically inexhaustible. These include solar, wind, geothermal, biomass, and wood resources. Also referred to as renewable energy.

Alternative fuel — Transportation fuels other than gasoline or diesel. Includes natural gas, methanol, and ethanol.

Articulate — A planned sequence of curriculum and course offerings from grades K-12.

Artifact — A human-made object.

Artificial ecosystem — Human-made environment or system that functions as a replication of or to produce the equivalent of the natural environment.

Assessment — 1. An evaluation technique for technology that requires analyzing benefits and risks, understanding the trade-offs, and then determining the best action to take in order to ensure that the desired positive outcomes outweigh the negative consequences. 2. An exercise, such as an activity, portfolio, written test, or experiment that seeks to measure a student's skills or knowledge in a subject area. Information may be collected about teacher and student performance, student behavior, and classroom atmosphere.

Batch production — The process of producing parts or components in quantity to be assembled into larger products.

Benchmark — 1. A written statement that describes the specific developmental components by various grade levels (K-2, 3-5, 6-8, and 9-12) that students should know or be able to do in order to achieve a standard. 2. A criteria by which something can be measured or judged.

Biodegradable — The ability of a substance to be broken down physically and/or chemically by natural biological processes, such as by being digested by bacteria or fungi.

Bioengineering — Engineering applied to biological and medical systems, such as biomechanics, biomaterials, and biosensors. Bioengineering also includes biomedical engineering as in the development of aids or replacements for defective or missing body organs.

Biological processes — The processes characteristic of, or resulting from, the activities of living organisms.

Biotechnology — Any technique that uses living organisms, or parts of organisms, to make or modify products, improve plants or animals, or to develop microorganisms for specific uses.

Brainstorming — A method of shared problem solving in which all members of a group spontaneously and in an unrestrained discussion generate ideas.

Bronze Age — The stage or level of development of human culture that followed the Stone Age and was characterized by the use of bronze tools and weapons and ended with the advent of the Iron Age; about 3000 B.C.E. to 1100 B.C.E.

Build — To make something by joining materials or components together into a composite whole.

By-product — Something produced in the making of something else; a secondary result; a side effect.

CAD (computer-aided design or computer-aided drafting) — 1. (Design) The use of a computer to assist in the process of designing a part, circuit, building, etc. 2. (Drafting) The use of a computer to assist in the process of creating, storing, retrieving, modifying, plotting, and communicating a technical drawing.

Capital — One of the basic resources used in a technological system. Capital (money) is the accumulated finances and goods devoted to the production of other goods.

Category — As used in this document, the large organizers for the study of technology. The categories are: The Nature of Technology, Technology and Society, Design, Abilities for a Technological World, and The Designed World.

Chemical technology — Any technological process that modifies, alters, or produces chemical substances, elements, or compounds.

Closed-loop system — A system that uses feedback from the output to control the input.

Cognitive knowledge — The level of understanding just beyond comprehension (basic understanding of meaning). This may include the application of rules, methods, concepts, principles, laws, and theories.

Combining — The joining of two or more materials by such processes as fastening, coating, and making composites.

Communication — The successful transmission of information through a common system of symbols, signs, behavior, speech, writing, or signals.

Communication system — A system that forms a link between a sender and a receiver making possible the exchange of information.

Complex system — A system consisting of interconnected or interwoven parts that interact in such a way as to produce a global output that cannot always be predicted.

Component — A part or element of a whole that can be separated from or attached to a system.

Compost — Substance composed mainly of partly decayed organic material, used to fertilize the soil and increase its humus content. Usually made from plant materials (e.g., grass clippings and leaves), manure, and soil, and can include chemical fertilizers and lime.

Computer — A machine for carrying out calculations and performing specified transformations on information, such as storing, sorting, correlating, retrieving and processing data.

Conditioning processes — Processes (using force, heat, cold, electricity, etc.) in which the internal structure of a material is changed to alter its properties to make it stronger, improve its function or appearance.

Consequence — An effect that naturally follows and is caused by a previous action or condition; referred to as an outcome.

Conservation — The preservation and protection of the environment and the wise use of natural resources.

Constraint — A limit to the design process. Constraints may be such things as appearance, funding, space, materials, and human capabilities.

Construction — The systematic act or process of building, erecting, or constructing buildings, roads, or other structures.

Control — An arrangement of chemical, electronic, electrical, and mechanical components that commands or directs the management of a system.

Control system — An assemblage of control apparatus coordinated to execute a planned set of actions.

Convention — A technique, practice, or procedure that is established by usage and widely accepted.

Creative thinking — The ability or power used to produce original thoughts and ideas based upon reasoning and judgment.

Credentialed teachers — Teachers who are licensed by a state department of education in a particular area of competence in order to be qualified to teach a particular subject or group of subjects.

Criterion — A desired specification (element or feature) of a product or system.

Critical thinking — The ability to acquire information, analyze and evaluate it, and reach a conclusion or answer by using logic and reasoning skills.

Culture — The beliefs, traditions, habits, and values controlling the behavior of the majority of the people in a social-ethnic group. These include the people's way of dealing with their problems of survival and existence as a continuing group.

Curriculum — The subject matter that teachers and students cover in their studies. It describes and specifies the methods, structure, organization, balance and presentation of the content.

Curriculum development — The process of planned development of curriculum pedagogy, instruction, and presentation modes.

Custom production — A type of production in which products are designed and built to meet the specific needs and wants of an individual.

Data — Raw facts and figures that can be used to draw a conclusion.

Data processing system — A system of computer hardware and software to carry out a specified computational task.

Decision making — The act of examining several possible behaviors and selecting from them the one most likely to accomplish the individual's or group's intention. Cognitive processes such as reasoning, planning, and judgment are involved.

Decode — To convert a coded message into understandable form using ordinary language.

Design — An iterative decision-making process that produces plans by which resources are converted into products or systems that meet human needs and wants or solve problems.

Design brief — A written plan that identifies a problem to be solved, its criteria, and its constraints. The design brief is used to encourage thinking of all aspects of a problem before attempting a solution.

Design principle — Design rules regarding rhythm, balance, proportion, variety, emphasis, and harmony, used to evaluate existing designs and guide the design process.

Design process — A systematic problem-solving strategy, with criteria and constraints, used to develop many possible solutions to solve a problem or satisfy human needs and wants and to winnow (narrow) down the possible solutions to one final choice.

Design proposal — A written plan of action for a solution to a proposed problem.

Develop — To change the form of something through a succession of states or stages, each of which is preparatory to the next. The successive changes are undertaken to improve the quality of or refine the resulting object or software.

Developmentally appropriate — Educational programs and methods that are intended to match the needs of students in the areas of cognition, physical activity, emotional growth, and social adjustment.

Diagnose — To determine, by analysis, the cause of a problem or the nature of something.

Discipline — A formal branch of knowledge or teaching (e.g., biology, geography, and engineering) that is systematically investigated, documented, and taught.

Drawing — A work produced by representing an object or outlining a figure, plan, or sketch by means of lines. A drawing is used to communicate ideas and provide direction for the production of a design.

Durable goods — An item that can be used for many years.

Economy — The system or range of economic activity, such as production, distribution, and consumption in a country, region, or community that manages domestic affairs and resources.

Educational technology — Using multimedia technologies or audiovisual aids as a tool to enhance the teaching and learning process.

Efficient — Operating or performing in an effective and competent manner with a minimum of wasted time, energy, or waste products.

Emergent — Occurring as a consequence.

Encode — To change a message into symbols or a form that can be transmitted by a communication system.

Energy — The ability to do work. Energy is one of the basic resources used by a technological system.

Engineer — A person who is trained in and uses technological and scientific knowledge to solve practical problems.

Engineering — The profession of or work performed by an engineer. Engineering involves the knowledge of the mathematical and natural sciences (biological and physical) gained by study, experience, and practice that are applied with judgment and creativity to develop ways to utilize the materials and forces of nature for the benefit of mankind.

Engineering design — The systematic and creative application of scientific and mathematical principles to practical ends such as the design, manufacture, and operation of efficient and economical structures, machines, processes, and systems.

Ergonomics — The study of workplace equipment design or how to arrange and design devices, machines, or workspace so that people and things interact safely and most efficiently. Also called human factors analysis or human factors engineering.

Ethical — Conforming to an established set of principles or accepted professional standards of conduct.

Evaluation — 1. The collection and processing of information and data in order to determine how well a design meets the requirements and to provide direction for improvements. 2. A process used to analyze, evaluate, and appraise a student's achievement, growth, and performance through the use of formal and informal tests and techniques.

Experimentation — 1. The act of conducting a controlled test or investigation. 2. The act of trying out a new procedure, idea, or activity.

Fact — A statement or piece of information that is true or a real occurrence.

Feedback — Using all or a portion of the information from the output of a system to regulate or control the processes or inputs in order to modify the output.

Figure — A written symbol, other than a letter, representing an item or relationship, especially a number, design, or graphic representation.

Forecast — A statement about future trends, usually as a probability, made by examining and analyzing available information. A forecast is also a prediction about how something will develop usually as a result of study and analysis of available pertinent data.

Forming — The process that changes the shape and size of a material without cutting it.

Grade-level — A stage in the development of a child's education; an acceptable grouping of different grades in school (e.g., K-2, 3-5, 6-8, and 9-12).

Guidance system — A system that provides information for guiding the path of a vehicle by means of built-in equipment and control.

Human factors engineering — See Ergonomics.

Human wants and needs — Human wants refers to something desired or dreamed of and human needs refers to something that is required or a necessity.

Hydroponics — A technique of growing plants without soil, in water or sometimes an inert medium (e.g., sand) containing dissolved nutrients.

Hypertext Markup Language (HTML) — The computer language used to create World Wide Web pages, with hyperlinks and markup for text formatting.

Impact — The effect or influence of one thing on another. Some impacts are anticipated, and others are unanticipated.

Industrial Revolution — A period of inventive activity, beginning around 1750 in Great Britain. During this period, industrial and technological changes resulted in mechanized machinery that replaced much of which was previously manual work. The Industrial Revolution was responsible for many social changes, as well as changes in the way things were manufactured.

Information — One of the basic resources used by technological systems. Information is data and facts that have been organized and communicated in a coherent and meaningful manner.

Information Age — A period of activity starting in the 1950s and continuing today in which the gathering, manipulation, classification, storage, and retrieval of information is central to the workings of society. Information is presented in various forms to a large population of the world through the use of machines, such as computers, facsimile machines, copiers, and CD-ROMs. The Information Age was enhanced by the development of the Internet; an electronic means to exchange information in short periods of time, often instantaneously.

Information system — A system of elements that receive and transfer information. This system may use different types of carriers, such as satellites, fiber optics, cables, and telephone lines, in which switching and storage devices are often important parts.

Infrastructure — 1. The basic framework or features of a system or organization. 2. The basic physical systems of a country's or a community's population, including transportation and utilities.

Innovation — An improvement of an existing technological product, system, or method of doing something.

Inorganic — Lacking the qualities, structure, and composition of living organisms; inanimate.

Input — Something put into a system, such as resources, in order to achieve a result.

In-service — 1. A full-time employee. 2. Workshops and lectures designed to keep practicing professionals abreast of the latest developments in their field.

Instructional technology — The use of computers, multimedia, and other technological tools to enhance the teaching and learning process. Sometimes referred to as educational technology.

Integration — The process of bringing all parts together into a whole.

Intelligence — The capacity to acquire knowledge and the skilled use of reason; the ability to comprehend.

Intelligent transportation system — Proposed evolution of the entire transportation system involving the use of information technologies and advances in electronics in order to revolutionize all aspects of the transportation network. These technologies include the use of the latest computers, electronics, communications, and safety systems to provide traffic control, freeway and incident management, and emergency response.

Interdisciplinary instruction — An educational approach where the students study a topic and its related issues in the context of various academic areas or disciplines.

Intermodalism — The use of more than one form of transportation.

Internet — The worldwide network of computer links, begun in the 1970s, which today allows computer users to connect with other computer users in nearly every country, and speaking many languages.

Invention — A new product, system, or process that has never existed before, created by study and experimentation.

Iron Age — The period of human culture characterized by the smelting of iron and its use in industry beginning after the Bronze Age somewhat before 1000 B.C. in western Asia and Egypt.

Irradiation — Treatment through the use of ionizing radiation, such as X-rays or radioactive sources (e.g., radioactive iodine seeds).

Irrigation system — A system that uses ditches, pipes, or streams to distribute water artificially.

Iterative — Describing a procedure or process that repeatedly executes a series of operations until some condition is satisfied. An iterative procedure may be implemented by a loop in a routine.

Just-in-Time (JIT) **manufacturing** — A systems approach to developing and operating a manufacturing system, in which manufacturing operation component parts arrive *just in time* to be picked up by a worker and used.

Kinetic energy — The energy possessed by a body as a result of its motion.

Knowledge — 1. The body of truth, information, and principles acquired by mankind. 2. Interpreted information that can be used.

Laboratory-classroom — The formal environment in school where the study of technology takes place. At the elementary school, this environment will likely be a regular classroom. At the middle and high school levels, a separate laboratory with areas for hands-on activities, as well as group instruction, could constitute the environment.

Literacy — Basic knowledge and abilities required to function adequately in one's immediate environment.

Machine — A device with fixed and moving parts that modifies mechanical energy in order to do work.

Maintenance — The work needed to keep something in proper condition; upkeep.

Management — The act of controlling production processes and ensuring that they operate efficiently and effectively; also used to direct the design, development, production, and marketing of a product or system.

Manufacturing — The process of making a raw material into a finished product; especially in large quantities.

Manufacturing system — A system or group of systems used in the manufacturing process to make products for an end user.

Market — 1. A subset of the population considered to be interested in the buying of goods or services. 2. A place where goods are offered for sale.

Marketing — The act or process of offering goods or services for sale.

Mass production — The manufacture of goods in large quantities by means of machines, standardized design and parts, and, often, assembly lines.

Material — The tangible substance (chemical, biological, or mixed) that goes into the makeup of a physical object. One of the basic resources used in a technological system.

Mathematics — The science of patterns and order and the study of measurement, properties, and the relationships of quantities; using numbers and symbols.

Measurement — The process of using dimensions, quantity, or capacity by comparison with a standard in order to mark off, apportion, lay out, or establish dimensions.

Medical technology — Of or relating to the study of medicine through the use of and advances of technology, such as medical instruments and apparatus, imaging systems in medicine, and mammography. Related terms: biomedical engineering and medical innovations.

Medicine — The science of diagnosing, treating, or preventing disease and other damage to the body or mind.

Message — 1. The information sent by one source to another, usually short and transmitted by words, signals, or other means. 2. An arbitrary amount of information whose beginning and end are defined or implied.

Micro-processing system — A computer made up of integrated circuits that is capable of high speed electronic operations.

Middle Ages — The period in European history between antiquity and the Renaissance, often dated from A.D. 476 to 1453.

Mixed-natural materials — Natural materials modified to improve their properties. Mixed-natural materials may be leather, plywood, or paper, for example.

Mobility — The quality or state of being mobile; capable of moving or being moved.

Model — A visual, mathematical, or three-dimensional representation in detail of an object or design, often smaller than the original. A model is often used to test ideas, make changes to a design, and to learn more about what would happen to a similar, real object.

Module — A self-contained unit.

Multimedia — Information that is mixed and transmitted from a number of formats (e.g., video, audio, and data).

Natural material — Material found in nature, such as wood, stone, gases, and clay.

Network — An interconnected group or system. The Internet is a network of computers.

Noise — An outside signal that interrupts, interferes, or reduces the clarity of a transmission.

Non-biodegradable — The inability of a substance to be broken down (decomposed) and therefore retaining its form for an extended period of time.

Non-durable goods — Items that do not last and are constantly consumed, such as paper products.

Nonlinear — Not in a straight line.

Nonrenewable — An object, thing, or resource that cannot be replaced.

Nuclear power — Power, the source of which is nuclear fission or fusion.

Obsolescence — Loss in the usefulness of a product or system because of the development of an improved or superior way of achieving the same goal.

Open-loop system — A control system that has no means for comparing the output with input for control purposes. Control of open-loop systems often requires human intervention.

Optimization — An act, process, or methodology used to make a design or system as effective or functional as possible within the given criteria and constraints.

Output — The results of the operation of any system.

People — One of the basic resources in a technological system. Humans design, develop, produce, use, manage, and assess products and systems.

Plan — A set of steps, procedures or programs, worked out beforehand in order to accomplish an objective or goal.

Political — Of or relating to the structure and affairs of a government, state, or locality and their related politics.

Pollution — The changing of a natural environment, either by natural or artificial means, so that the environment becomes harmful or unfit for living things; especially applicable to the contamination of soil, water, or the atmosphere by the discharge of harmful substances.

Portfolio — A systematic and organized collection of a student's work that includes results of research, successful and less successful ideas, notes on procedures, and data collected.

Potential energy — The energy of a particle, body, or system that is determined by its position or structure.

Power — 1. The amount of work done in a given period of time. 2. The source of energy or motive force by which a physical system or machine is operated.

Power system — A technological system that transforms energy resources to power.

Pre-service — Undergraduate coursework taken by those intending to teach.

Problem solving — The process of understanding a problem, devising a plan, carrying out the plan, and evaluating the plan in order to solve a problem or meet a need or want.

Procedural knowledge — Knowing how to do something.

Process — 1. Human activities used to create, invent, design, transform, produce, control, maintain, and use products or systems; 2. A systematic sequence of actions that combines resources to produce an output.

Produce — To create, develop, manufacture, or construct a human-made product.

Product — A tangible artifact produced by means of either human or mechanical work, or by biological or chemical processes.

Product lifecycle — Stages a product goes through from concept and use to eventual withdrawal from the marketplace. Product life cycle stages include research and development, introduction, market development, exploitation, maturation, saturation, and finally decline.

Production system — A technological system that involves producing products and systems by manufacturing (on the assembly line) and construction (on the job).

Propulsion system — A system that provides the energy source, conversion, and transmission of power to move a vehicle.

Prototype — A full-scale working model used to test a design concept by making actual observations and necessary adjustments.

Quality control — A system by which a desired standard of quality in a product or process is maintained. Quality control usually requires feeding back information about measured defects to further improvements of the process.

Receiver — The part of a communication system that picks up or accepts a signal or message from a channel and converts it to perceptible forms.

Recycle — To reclaim or reuse old materials in order to make new products.

Renaissance — The transitional movement in Europe between medieval and modern times beginning in the 14th century in Italy, lasting into the 17th century, and marked by a humanistic revival of classical influence expressed in a flowering of the arts and literature and by the beginnings of modern science.

Renewable — Designation of a commodity or resource, such as solar energy or firewood, that is inexhaustible or capable of being replaced by natural ecological cycles or sound management practices.

Requirements — The parameters placed on the development of a product or system. The requirements include the safety needs, the physical laws that will limit the development of an idea, the available resources, the cultural norms, and the use of criteria and constraints.

Research and development (R&D) — The practical application of scientific and engineering knowledge for discovering new knowledge about products, processes, and services, and then applying that knowledge to create new and improved products, processes, and services that fill market needs.

Resource — The things needed to get a job done. In a technological system, the basic technological resources are: energy, capital, information, machines and tools, materials, people, and time.

Risk — The chance or probability of loss, harm, failure, or danger.

Sanitation — The design and practice of methods for solving basic public health problems, such as drainage, water and sewage treatment, and waste removal.

Scale — A proportion between two sets of dimensions used in developing accurate, larger or smaller prototypes or models of design ideas.

Schematic — A drawing or diagram of a chemical, electrical, or mechanical system.

Science — The study of the natural world through observation, identification, description, experimental investigation, and theoretical explanations.

Scientific inquiry — The use of questioning and close examination using the methodology of science.

Sender — A person or equipment that causes a message to be transmitted.

Separating — The process of using machines or tools to divide materials.

Service — 1. The installation, maintenance, or repairs provided or completed by a dealer, manufacturer, owner, or contractor. 2.The performance of labor for the benefit of another.

Side effect — A peripheral or secondary effect, especially an undesirable secondary effect. Some side effects become the central basis for new developments.

Sketch — A rough drawing representing the main features of an object or scene and often made as a preliminary study.

Skill — An ability that has been acquired by training or experience.

Society — A community, nation, or broad grouping of people having common traditions, institutions, and collective activities and interests.

Solution — A method or process for solving a problem.

Standard stock item — A supply of items that are commonly used and kept in inventory for quick access.

Standardization — The act of checking or adjusting by comparison with a standard.

Stone Age — The first known period of prehistoric human culture characterized by the use of stone tools.

Structural system — A system comprised of the framework or basic structure of a vehicle.

Structure — Something that has been constructed or built of many parts and held or put together in a particular way.

Subsystem — A division of a system that, in itself, has the characteristics of a system.

Support system — 1. A network of personnel or professionals that provides life, legal, operational, maintenance, and economic support for the safe and efficient operation of a system, such as a transportation system. 2. The technical system that supports the operation of a system, as in a life support system on board the Shuttle.

Suspension system — A system of springs and other devices that insulates the passenger compartment of a vehicle from shocks transmitted by the wheels and axles.

Sustainable — 1. Of, relating to, or being a method of harvesting or using a resource so that the resource is not depleted or permanently damaged. 2. Relating to a human activity that can be sustained over the long term, without adversely affecting the environmental conditions (soil conditions, water quality, climate) necessary to support those same activities in the future.

Symbol — An arbitrary or conventional sign that is used to represent operations, quantities, elements, relations, or qualities or to provide directions or alert one to safety.

Synthetic material — Material that is not found in nature, such as glass, concrete, and plastics.

System — A group of interacting, interrelated, or interdependent elements or parts that function together as a whole to accomplish a goal.

Systems-oriented thinking — A technique for looking at a problem in its entirety, looking at the whole, as distinct from each of its parts or components. Systems-oriented thinking takes into account all of the variables and relates social and technological characteristics.

Technological design — See Engineering design.

Technological literacy — The ability to use, manage, understand, and assess technology.

Technology — 1. Human innovation in action that involves the generation of knowledge and processes to develop systems that solve problems and extend human capabilities. 2. The innovation, change, or modification of the natural environment to satisfy perceived human needs and wants.

Technology content standard — A written statement that specifies the knowledge (what students should know) and process (what students should be able to do) students should possess in order to be technologically literate.

Technology education — A study of technology, which provides an opportunity for students to learn about the processes and knowledge related to technology that are needed to solve problems and extend human capabilities.

Technological studies — See Technology education.

Technological transfer — The process by which products, systems, knowledge, or skills, developed under federal research and development funding, is translated into commercial products to fulfill public and private needs.

Telemedicine — The investigation, monitoring, and management of patients and the education of patients and staff using systems which allow ready access to expert advice and patient information, no matter where the patient or the relevant information is located. The three main dimensions of telemedicine are health service, telecommunications, and medical computer technology.

Test — 1. A method for collecting data. 2. A procedure for critical evaluation.

Thematic unit — Set of lesson presentations that organize classroom instruction around certain texts, activities, and learning episodes related to a topic(s). A thematic unit might integrate several content areas.

Tool — A device that is used by humans to complete a task.

Trade-off — An exchange of one thing in return for another; especially relinquishment of one benefit or advantage for another regarded as more desirable.

Transmit — To send or convey a coded or non-coded message from a source to a destination.

Transportation system — The process by which passengers or goods are moved or delivered from one place to another.

Trend — 1. A tendency; 2. A general direction.

Trend analysis — A comparative study of the component parts of a product or system and the tendency of a product or system to develop in a general direction over time.

Trial and error — A method of solving problems in which many solutions are tried until errors are reduced or minimized.

Troubleshoot — To locate and find the cause of problems related to technological products or systems.

Use — The act or practice of employing something to put it into action or service.

Vignette — A brief description or verbal snapshot of how a standard or group of standards may be implemented in the laboratory-classroom.

Virtual — Simulation of the real thing in such a way that it presents reality in essence or in effect though not in actual fact.

Vocational education — Training within an educational institution that is intended to prepare an individual for a particular career or job.

Waste — Refuse or by-products that is perceived as useless, and must be consumed, left over, or thrown away.

Work — The transfer of energy from one physical system to another expressed as the product of a force and the distance through which it moves a body in the direction of that force

World Wide Web (WWW) — An abstract (imaginary) space of information, which includes documents, color images, sound, and video.

Index